Science Ideated

The fall of matter and the contours of the next mainstream scientific worldview

Science Ideated

The fall of matter and the contours of the next mainstream scientific worldview

Bernardo Kastrup

IFF
BOOKS

Winchester, UK
Washington, USA

JOHN HUNT PUBLISHING

First published by iff Books, 2021
iff Books is an imprint of John Hunt Publishing Ltd., No. 3 East Street, Alresford,
Hampshire SO24 9EE, UK
office@jhpbooks.com
www.johnhuntpublishing.com
www.iff-books.com

For distributor details and how to order please visit the 'Ordering' section on our website.

ISBN: 978 1 78904 668 7
978 1 78904 669 4 (ebook)
Library of Congress Control Number: 2020938259

A CIP catalogue record for this book is available from the British Library.

Design: Stuart Davies

UK: Printed and bound by CPI Group (UK) Ltd, Croydon, CR0 4YY
Printed in North America by CPI GPS partners

Contents

Other Books by Bernardo Kastrup

Rationalist Spirituality: An exploration of the meaning of life and existence informed by logic and science

Dreamed up Reality: Diving into mind to uncover the astonishing hidden tale of nature

Meaning in Absurdity: What bizarre phenomena can tell us about the nature of reality

Why Materialism Is Baloney: How true skeptics know there is no death and fathom answers to life, the universe, and everything

Brief Peeks Beyond: Critical essays on metaphysics, neuroscience, free will, skepticism and culture

More Than Allegory: On religious myth, truth and belief

The Idea of the World: A multi-disciplinary argument for the mental nature of reality

Decoding Schopenhauer's Metaphysics: The key to understanding how it solves the hard problem of consciousness and the paradoxes of quantum mechanics

Decoding Jung's Metaphysics: The archetypal semantics of an experiential universe

Brief Introduction

The story of how science and metaphysical materialism became seemingly intertwined is a curious one. Back in the seventeenth century, when science as we know it today took its first steps, scientists based their entire work on—what else?—*perceptual experience*: the things and phenomena they could see, touch, smell, taste or hear around them. That starting point is, of course, *qualitative* in nature. After all, the felt concreteness of the proverbial apple that fell on Newton's head, as well as its redness and sweetness, were *sensed qualities*. Everything that appears on the screen of perception is perforce qualitative. As such, the starting point of science—then and now—is the world of qualities that we perceive around ourselves. Even the output of perception-enhancing instruments such as microscopes and telescopes is only useful insofar as it is qualitatively perceived.

Soon, however, scientists realized that it is very convenient to *describe* this eminently qualitative world by means of *quantities,* such as weights, lengths, angles, speeds, etc. These quantities capture the relative differences between qualities. For instance, an anvil feels qualitatively heavier than a feather, a difference in felt weight that can be conveniently described with a quantity: a certain number of newtons. Today we have units—quantities—to describe every discernible aspect of the world, including frequency, amplitude, mass, charge, momentum, spin, etc.

But then something bizarre happened: many scientists seemingly forgot where it all started and began attributing fundamental reality *only to the quantities*. Because only quantities can be objectively measured, they began postulating that only mass, charge, momentum, etc., really exist out there, qualities somehow being ephemeral epiphenomena—side effects—of brain activity, existing only within the confines of our skull. This, in a nutshell, was the birth of metaphysical materialism, a philosophy that—absurdly—grants fundamental reality to mere *descriptions*, while denying the reality of that which is described in the first place.

Indeed, at some point between the early seventeenth and the late nineteenth century, we began cluelessly replacing reality with its description, the territory with the map. Now we say that only matter exists—i.e. things exhaustively defined in terms of quantities alone—while the *qualities of experience*, which are all we ultimately have, are allegedly secondary, reducible, epiphenomenal. And so we now face the so-called 'hard problem of consciousness': the impossibility of explaining qualities in terms of quantities. That we find ourselves surprised at the intractability of this 'problem' is what is dumbfounding: we *defined* matter as something *purely quantitative*—i.e. *not* a quality—to begin with, so it's no wonder that we can't reduce qualities to matter, is it?

The hope that we will one day solve the 'hard problem' is as foolish as hoping to reduce the territory to its map, a painter to his or her self-portrait. The hard problem must be *seen through* and *circumvented*, not solved. Our present metaphysical dilemmas—as well as the story that brought us to them, as briefly outlined above—would be comical if they weren't tragic. In the space of only a couple of centuries, we tied ourselves up in hopelessly abstract conceptual knots and managed to lose touch with reality altogether.

If science is to progress beyond its present dilemmas—from those in the neuroscience of consciousness to those in the foundations of quantum mechanics, which have their roots in the same conceptual misstep described above—we must undo the knots and place our feet back on firm ground. This book is an attempt to help us do just that.

Indeed, leading-edge empirical observations are increasingly difficult to reconcile with metaphysical materialism. Laboratory results in quantum mechanics, for instance, strongly indicate that there is no autonomous material world of tables and chairs out there. Coupled with the inability of materialist neuroscience to explain experience, this is finally forcing us to reexamine our early assumptions and contemplate alternatives. *Analytic idealism*—the notion that reality, while equally amenable to scientific inquiry, is fundamentally qualitative—is a leading contender to replace metaphysical materialism.

In this book, the broad body of empirical evidence and reasoning in favor of analytic idealism is reviewed in an accessible manner. The book consists of a compact collection of essays written between 2017 and 2020. The original versions of most of them have previously been published in preeminent magazines and journals—such as *Scientific American*, the *Journal of Near-Death Studies*, *IAI News* (the online magazine of the *Institute of Art and Ideas*), the *Blog of the American Philosophical Association* and *Science and Nonduality*—as well as my own blog. They are collected here in a convenient format, ordered and grouped together in a manner that facilitates their understanding.

The essays have been revised, updated and sometimes extended. Often the original versions had to comply with editorial preferences not my own, whereas the versions in this book are my preferred ones: the 'director's cut,' so to speak, reflecting my true tone and style. Two never-before-published essays are also included: *Why Does Nature Mirror Our Reasoning?* (Chapter 23) and *Is Life More Than Physics?* (Chapter 24).

The essays often—though not always—address subjects previously covered in earlier books of mine. However, they embody an increased clarity of argument developed since then. As such, the present book is my chance to cover old ground in a new, fresh way, sharper and more concise. In a sense, it is a grand summary of my ideas: each chapter contains a short and focused distillation of at least one of the defining thoughts behind analytic idealism. The resulting argument anticipates a historically imminent transition to a scientific worldview that, while elegantly accommodating all known empirical evidence and predictive models, regards *mind*—not matter—as the ground of all reality.

More than any previous book of mine, this one includes criticisms of metaphysical materialism, consciousness denialism, panpsychism and other philosophical and scientific views prevalent in our culture today. In a sense, it is a concentrated, blazing reproach—no punches pulled—of the insanity that characterizes our mainstream worldview at the present historical juncture. This reproach is issued in the hope that it may help us change our most dysfunctional ways, so we can live closer to truth.

Part I

On 'Scientific' Materialism

Chapter 1

Why Materialism Is a Dead-End

How misunderstanding matter has led us astray

(The original version of this essay was published on IAI News *on 15 November 2019)*

We live in an age of science, which has enabled technological advancements unimaginable to our ancestors. Unlike philosophy, which depends somewhat on certain subjective values and one's own sense of plausibility to settle questions, science poses questions directly to nature, in the form of experiments. Nature then answers by displaying certain behaviors, so questions can be settled objectively.

This is both science's strength and its Achilles' heel: experiments only tell us how nature *behaves*, not what it essentially *is*. Many different hypotheses about nature's essence are consistent with its manifest behaviors. So although such behaviors are informative, they can't settle questions of *being*, which philosophers call 'metaphysics.' Understanding nature's essence is fundamentally beyond the scientific method, which leaves us with the—different—methods of philosophy. These, somewhat subjective as they may be, are our only path to figuring out what is going on.

'Scientific' materialism—the view that nature is fundamentally constituted by matter outside and independent of mind—is a *metaphysics*, in that it makes statements about what nature essentially *is*. As such, it is also a theoretical inference: we cannot empirically observe matter outside and independent of mind, for we are forever locked in mind. All we can observe are the contents of perception, which are inherently *mental*. Even the output of measurement instruments is only accessible to us insofar as it is mentally perceived.

We infer the existence of something beyond mental states because, at first, this seems to make sense of three canonical observations:

(i) We all seem to share the same world beyond ourselves.
(ii) The behaviour of this shared world doesn't seem to depend on our volition.
(iii) There are tight correlations between our inner experiences and measurable patterns of brain activity.

A world outside mental states, which we all inhabit, tentatively makes sense of observation (i). Because this shared world is thus *non*-mental, it isn't acquiescent to our (mental) volition, thereby tentatively explaining (ii). Finally, if particular configurations of matter in this world somehow generate mentality, it could perhaps also explain (iii). And so our culture has come to take for granted that nature is essentially material, non-mental. Again, this is a *metaphysical inference* aimed at tentatively explaining the canonical observations listed above, not a scientific or empirical fact.

The problem is that such metaphysical inference is untenable on several grounds. For starters, there is nothing about the parameters of material arrangements—say, the position and momentum of the atoms constituting our brain—in terms of which we could deduce, at least in principle, how it feels to fall in love, to taste wine or to listen to a Vivaldi sonata. There is an impassable explanatory gap between material *quantities* and experiential *qualities*, which philosophers refer to as the 'hard problem of consciousness' (Chalmers 2003). Many people don't recognize this gap because they think of matter as already having intrinsic qualities—such as color, taste, etc.—which contradicts 'scientific' materialism: according to the latter, color, taste, etc., are all generated by our brain, inside our skull. They don't exist in the world out there, which is supposedly purely abstract (see Chapter 3 of this book).

Second, materialism lives or dies with what physicists call 'physical realism': there must be an objective physical world out there, consisting of entities with defined properties, whether such world is being observed or not. The problem is that experiments over the past four decades have now refuted physical realism beyond reasonable doubt (see Chapters 16, 17, 20 and 21 of this book). So unless one redefines the meaning of the word 'materialism' in a rather arbitrary manner, 'scientific' materialism is now *physically* untenable.

Third, a compelling case can be made that the empirical data we have now amassed on the correlations between brain activity and inner experience cannot be accommodated by materialism. There is a broad, consistent pattern associating impairment or reduction of brain metabolism with an expansion of awareness, an enrichment of experiential contents and their felt intensity (see Chapter 25 of this book). It is at least difficult to see how the materialist hypothesis that all experiences are somehow generated by brain metabolism could make sense of this.

Finally, from a philosophical perspective, materialism is at least unparsimonious—that is, uneconomical, unnecessarily extravagant—and arguably even incoherent. Coherence and parsimony are admittedly somewhat subjective values. However, if we were to abandon them, we would have to open the gates to all kinds of nonsense: from aliens in the Pleiades trying to alert us to global catastrophe to teapots in the orbit of Saturn—neither of which can be empirically disproven. So we better stick to these values, for the price of having to apply them *consistently*, even to materialism itself.

Materialism is unparsimonious because, in addition to or instead of mentality—which is all we are directly acquainted with and ultimately know—it posits another category of 'substance' or 'existence' fundamentally beyond direct empirical verification: namely, matter. Under materialism, matter is literally transcendent, more inaccessible than any ostensive spiritual world posited by the world's religions. This would only be justifiable if there were no way of making sense of the three canonical observations listed earlier on the basis of mind alone; *but there is*.

Materialism conflates the need to posit something outside our *personal* minds with having to posit something outside *mind as a category*. All three observations can be made sense of if we postulate a *trans*personal field of mentation beyond our *personal* psyches (see Part IV of this book). As such, there is indeed a world out there, beyond us, which we all inhabit; but this world is *mental*, just as we are intrinsically mental agents. Seeing things this way completely circumvents the 'hard problem of consciousness,' as we no longer need to bridge the impassable gap between mind and non-mind, quality and quantity: everything is now mental, qualitative, perception consisting solely in a modulation of one (personal) set of qualities by another (transpersonal) set of qualities. We know this isn't a problem because it happens every day: our own thoughts and emotions, despite being qualitatively different, modulate one another all the time.

Finally, materialism is arguably incoherent. As we have seen, matter is a theoretical abstraction in and of mind. So when materialists try to reduce mind to matter, they are effectively trying to reduce mind to one of mind's own conceptual creations (Kastrup 2018b). This is akin to a dog chasing its own tail. Better yet, it is like a painter who, having painted a self-portrait, points at it and proclaims himself to *be* the portrait. The ill-fated painter then has to explain his entire conscious inner life in terms of patterns of pigment distribution on canvas. Absurd as this sounds, it is very much analogous to the situation materialists find themselves in.

The popularity of materialism is founded on a confusion:

somehow, our culture has come to associate it with science and technology, both of which have been stupendously successful over the past three centuries. But that success isn't attributable to materialism; it is attributable, instead, to our ability to inquire into, model and then predict nature's *behavior*. Science and technology could have been done equally well—perhaps even better—without any metaphysical commitment, or with another metaphysics consistent with such behavior. Materialism is, at best, an illegitimate hitchhiker, perhaps even a parasite, in that it preys on the psychology of those who do science and technology (Kastrup 2016d).

Indeed, in order to relate daily to nature, human beings need to tell themselves a story about what nature is. It is psychologically very difficult to remain truly agnostic regarding metaphysics, particularly when one is doing experiments. Even when this internal story is subliminal, it is still running like a basic operating system. And so it happens that materialism, because of its vulgar intuitiveness and naïve superficiality, offers a cheap and easy option for such inner storytelling. In addition, it has arguably also enabled early scientists and scholars to preserve a sense of meaning at a time when religion was losing its grip on our culture (*Ibid.*).

But now, in the 21st century, we can surely do better than that. We are now in a position to examine our hidden assumptions honestly, confront the evidence objectively, bring our own psychological needs and prejudices to the light of self-reflection, and then ask ourselves: Does materialism really add up to anything? The answer should be obvious: it just doesn't. Materialism is a relic from an older, naiver and less sophisticated age, when it helped investigators separate themselves from what they were investigating. But it has no place in this day and age.

Neither do we lack options, as we can now make sense of all canonical observations on the basis of mental states alone (Kastrup 2019, Part IV of this book). This constitutes a more persuasive, parsimonious and coherent alternative to materialism, which can also accommodate the available evidence better. The fundamentals of this alternative have been known at least since the early 19th century (Kastrup 2020); arguably even millennia earlier. It is entirely up to us today to explore it and, frankly, get our act together when it comes to metaphysics. We should know better than to—bizarrely—keep on embracing the untenable.

Chapter 2

Ignorance

The surprising thing materialism has going for it

(The original version of this essay was published on my blog, Metaphysical Speculations, *on 26 January 2020)*

There is a strange feeling I get every now and then: when some conclusion I had earlier drawn through thought is confirmed by direct observation, I often get the feeling that I, in fact, hadn't really appreciated the true force and implications of the conclusion; at least not as assuredly and vividly as when the confirmation comes. At that moment, the conclusion feels so much truer that, whatever reasons I had to believe it before, seem hazy in comparison. I think to myself, "I thought I knew this, but only now do I *really* know it."

This has happened to me a couple of times over the past weeks, as I found myself doing an exposé of eliminativism and illusionism—the ridiculous notions that consciousness doesn't exist (see Part II of this book). More specifically, I sought to refute the incoherent arguments of neuroscientist Michael Graziano and philosopher Keith Frankish (see Chapter 8 of this book). It was when Graziano attempted to reply to my criticisms (2020) that I got the strange feeling I tried to describe above: I thought to myself, "this guy just doesn't know what consciousness is! He doesn't have the capacity to introspect and self-reflect enough to recognize his own raw awareness."

A part of me fully expected the kind of reaction I got from Graziano: conceptual obfuscation, hand-waving, lack of substantive argumentation and failure to address the points in contention. Here is someone who denies that consciousness exists, so what else could one reasonably expect? But another part of me was very sincerely baffled, surprised by the living confirmation of what had been for me, up until that point, more like a conclusion from thought than direct experience. I mean, it's one thing to know rationally that the emperor must have no clothes; but it's another thing entirely to *see* the emperor standing naked right in front of you and think, "This is *really* happening."

Graziano is a Princeton neuroscientist, mind you; a Princeton neuroscientist who recently made the cover of *New Scientist* magazine (Graziano 2019). And he doesn't seem able to meta-cognize his own awareness; doesn't seem to understand what the 'hard problem of consciousness' (Chalmers 2003) is all about and why it is unavoidable under materialist premises. Not only that, he is a Princeton neuroscientist who couldn't even weave a conceptually consistent counterargument in his 'reply' of little more than 800 words (Graziano 2020). To me this is outright scary. Our emperors are parading naked—yet proudly—in front of us. Watch carefully, ignore the posturing cacophony around you, and you shall see it in horror.

The whole thing made me think of two old friends of mine, with whom I now have—unfortunately—little contact. Both are hardcore computer scientists: they were my colleagues many years ago. Both are very competent and knowledgeable in what they do. One is also very erudite when it comes to the arts and the classics. In summary, two highly intelligent and educated human beings. Yet, both are self-declared hardcore materialists. Both—just like Graziano—consider any non-materialist metaphysical position mystical woo.

This has always puzzled me. Only over the years did I slowly begin to realize how two otherwise intelligent people can be so biased against much more reasonable metaphysics: the problem is not that they don't understand these other metaphysics; the problem is that *they don't understand materialism*.

Once I made a passing comment to one of my friends, about

the eliminativist idea that the brain deceives itself into thinking it is conscious (never mind the fact that, if this were so, the deception would itself be conscious, and thus the argument would immediately implode). My friend looked at me with wide-open eyes, as if he had just had an 'Aha' moment, and said: "Yes! Of course! This must be it!" Here was an idea he deeply *wanted* to believe. "Don't you see the elegance of this explanation?" he continued. He had finally found a way to circumvent what he couldn't make sense of: the origin of consciousness from matter.

I just stared at my friend in disbelief, and then had a sudden insight: "He doesn't know what consciousness is ..." I thought. But then, immediately, a deeper insight: "No, it's more than that: what he doesn't know is what *matter* is supposed to be!" It became clear to me that, each time I said 'consciousness,' my friend associated the word not with his *experience* of hearing me say it, but instead with some private conceptual abstraction of his own mind. For him, the abstraction was so self-evident that it went completely unexamined; he couldn't even recognize it as an abstraction. And so it was impossible to continue the conversation.

Not that long ago, I was talking to my other friend while having a beer with him in my backyard. The conversation had drifted to metaphysics and I asked him: "Isn't it strange to think that, according to materialism, all this [I pointed to the flowers, bees and trees around us] is created inside our skull?" The reference, of course, was to the qualities of experience—such as colors and smells—which materialism says are created by our brains and don't exist in the world beyond our skull. He paused and looked at me as if I had just said something unholy and totally incomprehensible. Finally, with obvious exasperation, he asked: "What the hell can you possibly mean by that? All this stuff [pointing to my backyard] is out there; obviously it's not just inside our heads." I tried to explain what I meant, but to no avail.

It was clear that, according to my friend's private, implicit 'materialism,' colors, melodies, flavors, textures, etc., *are all really out there*; there is nothing else the objective world could consist of. I suspect he implicitly believes the brain creates only thoughts and emotions, not the qualities of perception. This, of course, not only deviates from any coherent formulation of materialism, it is a metaphysical contradiction in and of itself (for a more elaborate explanation of this claim, see Chapter 3 of this book).

And so I finally come to my point: I think the strongest thing materialism has going for it is that most materialists *do not actually understand or recognize what materialism entails and implies*. Materialism is so blatantly absurd that most casual materialists—I strongly suspect—replace it with one or another private, implicit *mis*apprehension of it in their own minds, which circumvents some of the absurdities at the price of internal contradictions conveniently overlooked. In other words, it is the naked *implausibility* of materialism that—ironically—makes it seem credible, for such implausibility forces many to unwittingly misinterpret materialism in whatever secret way seems to make sense to them.

Compounding the problem, many people—even otherwise intelligent ones—don't appear able to recognize the nature of their own raw awareness through self-reflective introspection. For this reason, they conflate matter with the qualities of experience, just as my friend thought of the colors, sounds and smells of my garden as the *thing in itself*, instead of mere phenomena produced by the brain. It is precisely this unexamined error that renders materialism plausible to them: they think the material world *is* the contents of perception, although materialism states unequivocally that it isn't.

I can forgive my friends: they are computer scientists, not philosophers or neuroscientists. They are also not picking up a megaphone and shouting to the world that consciousness doesn't exist; their views are their own; they aren't interested in preaching. But when it comes to Graziano and Frankish, things are different. They want to convince you that you are not *really* conscious. Their message is toxic, not only because it is nonsensical, but because—if believed—it could undermine the very foundations of our secular ethics and moral codes. After all, if you weren't really conscious, you couldn't really suffer or feel real pain, could you? Do you see the danger of this nonsense ever becoming widely believed? So I, for one, will persist in pointing at them and shouting as loud as I can: "Look! They have no clothes!"

Chapter 3

A Materialism of Qualities?

Dispelling a popular misinterpretation of materialism

(The original version of this essay was published on my blog, Metaphysical Speculations, *on 29 January 2020)*

In the previous chapter, I suggested that some people who proclaim to adhere to the materialist metaphysics in fact misapprehend what materialism is. One example of misapprehension I mentioned was the implicit notion that, although the brain produces the felt *qualities* we call thoughts and emotions—that is, endogenous experiences—the qualities of perception, such as color, flavor, smell, etc., are thought to exist out there in the world, not inside our skull. These people subliminally assume that the physical world is constituted by the qualities displayed on the screen of perception, which contradicts 'scientific' materialism.

Indeed, according to materialism all qualities, *including those of perception,* are somehow—materialists don't know how—generated by the brain inside our skull. The external world allegedly has no qualities at all—no color, no smell, no flavor—but is instead constituted by purely abstract *quantities,* such as mass, charge, spin, momentum, geometric relationships, frequencies, amplitudes, etc. If these quantities were to be fully specified in the context of the mathematical equations underlying our physics, 'scientific' materialism maintains that *nothing else* would need to be said about the world; the quantities alone would allegedly define it *completely*.

But could we—for the sake of curiosity—conceive of an *alternative* but coherent form of materialism that acquiesced to the misinterpretation discussed above? That is, could we devise a coherent 'qualitative materialism' according to which the qualities of perception are really out there in the external world—whether they fully constitute that world or are merely objective properties of it—while only non-perceptual experiences, such as thoughts and emotions, are generated by the brain? The answer is most definitely 'no.'

For starters, notice that the qualities of perception—color, smell, flavor, etc.—also appear in dreams, imagination, visions, hallucinations, etc. Many dreams and hallucinations are qualitatively indistinguishable from actual perceptions, something I have verified multiple times—to my own satisfaction—during lucid dreams and psychedelic trances. So if colors and other perceptual qualities are really out there in the external world, then somehow our inner mental imagery can also incorporate the *exact same* qualities *independently* of the external world.

This is problematic, for it entails postulating two fundamentally different grounds for the same qualities: in one case, the qualities are inherent to the matter out there in the world; in the other case, the exact same qualities are somehow generated by material arrangements in our brain, which themselves, *ex hypothesi*, do *not* have those qualities.

For instance, the brain—that reddish-greyish object inside our skull—does not itself display the colors of the rainbow when we look at it on an operating table. Yet it obviously can generate the dream-imagery of a rainbow. Analogously, the brain itself does not sound like anything. Yet it can generate the dream of a lovely concert. So the *same* qualities must be *both* intrinsic to matter when they occur outside our skull, *and* also epiphenomena of material arrangements when they occur inside. This doesn't seem coherent to me.

You see, even if the perceptual qualities of our inner mental imagery are just remembered from earlier perceptions, under qualitative materialism the brain still has to epiphenomenally generate the experience of *re-living the memories*, despite not having the entailed qualities in its own matter. For instance, the brain has to epiphenomenally generate the re-experiencing of a rainbow—which entails experiencing many colors—without having all those colors in its own matter. So we still end up with two fundamentally different grounds for the same qualities.

But that's not all. The defining principle of all formulations of metaphysical materialism is that the classical, macroscopic world beyond our private mentation, *as it is in itself*, is objective; that is, its properties are independent of observation. Under qualitative materialism, this means that the perceptual qualities of an object—such as e.g. its color—are objective, intrinsic to the object itself, not private creations of our personal mind. Therefore, *these qualities can only change if the object itself changes*.

But visual illusions immediately disprove this. For instance, in the well-known 'checker shadow' illusion created by the Perceptual Science Group of the Massachusetts Institute of Technology, two identically colored squares—*A* and *B*—of a checkerboard are initially perceived to have different colors—dark and light grey, respectively—because of the different contexts in which they are perceived. But by looking at square *B* as it is moved towards square *A*, one sees that they indeed have the same color: even though the squares themselves aren't changed at all, the light grey seems to vanish, only dark grey being perceived in both squares. What makes this particular illusion so compelling is its robustness: even if you know in advance that it is an illusion, you just can't help but still *see* light grey in square *B*, before it is moved.

Under qualitative materialism, the perceived color of a square is *intrinsic to the square itself*; it is objective, existing beyond our personal mentation; square *B*, as it is in itself, *is* light grey. Therefore, for as long as we don't change anything about the square, its color should remain unchanged. But the illusion proves that such is not the case, in that the perceived color *does* change, whereas the square itself doesn't. If by altering merely what is going on *around* squares *A* and *B* we manage to make a color disappear, how could this color—this perceived quality—exist 'out there,' beyond our personal mind, to begin with? How could it be objective?

Mainstream 'scientific' materialism preserves the objectivity of the classical, macroscopic world around us by stating that the colors—or any other quality, for that matter—we perceive are generated by our brain, inside our skull. This internal generation of qualities depends not only on the internal characteristics of our visual system, but also on the external context of observation. This is why, according to mainstream materialism, we perceive the colors of the squares differently depending on context. Hence, visual illusions do not contradict mainstream materialism and are not the reason why it fails.

Qualitative materialism, on the other hand, has problems accommodating not only color illusions, but *any* perceptual illusion. It is also incoherent in that it requires two fundamentally different grounds for the same experiential qualities: one irreducible and the other epiphenomenal. Self-declared materialists who unwittingly associate the plausibility of their position with this misapprehension of what materialism means should thus rush to revise their worldview.

Chapter 4

Consciousness Cannot Have Evolved

Under 'scientific' materialism, the felt qualities of experience have no survival function

(The original version of this essay was published on IAI News *on 5 February 2020)*

The overwhelmingly validated theory of evolution tells us that the functions performed by our organs arose from associated increases in survival fitness. For instance, the bile produced by our liver and the insulin produced by our pancreas help us absorb nutrients and thus survive. Insofar as it is produced by the brain, our phenomenal consciousness—i.e. our ability to subjectively experience the world and ourselves—is no exception: it, too, must give us some survival advantage, otherwise natural selection wouldn't have fixed it in our genome. In other words, our sentience—to the extent that it is produced by the brain—must perform a beneficial function, otherwise we would be unconscious zombies.

The problem is that, under the premises of 'scientific' materialism, phenomenal consciousness cannot—*by definition*—have a function. Indeed, according to materialism all entities are defined and exhaustively characterized in purely *quantitative* terms. For instance, elementary subatomic particles are exhaustively characterized in terms of e.g. mass, charge and spin values. Similarly, the behavior of abstract fields is fully defined in terms of quantities, such as frequencies and amplitudes of oscillation. Particles and fields, in and of themselves, have *quantitative* properties but no intrinsic *qualities*, such as color or flavor. Only our perceptions of them—or so the materialist argument goes—are accompanied by qualities somehow generated by our brain.

Still under materialism, the quantities that characterize physical entities are what allows them to be causally efficacious; that is, to produce effects. For instance, it is the charge values of protons and electrons that produce the effect of their mutual attraction. In nuclear fission reactors, it is the mass value of neutrons that produces the effect of splitting atoms. And so on. All chains of cause and effect in nature must be describable purely in terms of quantities, for only quantities figure in the mathematical equations underlying physical theory. Whatever isn't a quantity cannot be part of our physical models and therefore—insofar as such models are presumed to be causally-closed—cannot produce effects. According to materialism, all functions rest on quantities.

However, our phenomenal consciousness is eminently *qualitative*, not quantitative. There is something it feels like to see the color red, which is not captured by merely noting the frequency of red light. If we were to tell someone born blind that red is an oscillation of approximately $4.3*10^{14}$ cycles per second, they would still not know what it feels like to see red. Analogously, what it feels like to listen to a Vivaldi sonata cannot be conveyed to a person born deaf, even if we show to the person the sonata's complete power spectrum. Experiences are felt qualities—which philosophers and neuroscientists call 'qualia'—not fully describable by abstract quantities.

As discussed above, qualities have no function under materialism, for quantitatively-defined physical models are supposed to be causally-closed; that is, sufficient to explain every natural phenomenon. As such, it must make no difference to the survival fitness of an organism whether the data processing taking place in its brain is accompanied by experience or not: whatever the case, the processing will produce the same effects; the organism will behave in exactly the same way and stand exactly the same chance to survive and reproduce. Qualia are, at best, superfluous extras.

Therefore, under physicalist premises phenomenal consciousness cannot have been favored by natural selection. Indeed, *it shouldn't exist at all*; we should all be unconscious zombies, going about our business in exactly the same way we actually do, but without accompanying inner life. If evolution is true—which we have every reason to believe is the case—our very sentience contradicts 'scientific' materialism.

This inescapable conclusion is often ignored by materialists, who regularly try to artificially attribute functions to phenomenal consciousness. Here are three illustrative examples:

(i) Consciousness enables *attention*.
(ii) Consciousness *discriminates* episodic memory (past) from live perceptions (present) by making them feel different.
(iii) Consciousness *motivates* behavior conducive to survival.

Computer scientists know that none of these require experience, for we routinely implement all three functions in presumably unconscious silicon computers.

Regarding point (i), under materialism attention is simply a mechanism for focusing an organism's limited cognitive resources on priority tasks. Computer operating systems do this all the time—using techniques such as interrupts, queuing, task scheduling, etc.—in a purely algorithmic, quantitatively-defined manner.

Regarding point (ii), there are countless ways to discriminate data streams without need for accompanying experience. Does your home computer have trouble separating the photos of last year's holidays from the live feed of your webcam? Data streams from memory and real-time processes can simply be tagged or routed in different ways, without qualia.

Finally, regarding point (iii), within the logic of materialism motivation is simply a calculation, the output of a quantitative algorithm tasked with maximizing the gain while minimizing the risk of an organism's actions. Computers are 'motivated' to do whatever it is they do—otherwise they wouldn't do it—without accompanying qualia.

Just as these three examples illustrate, all conceivable cognitive functions can, under materialist premises, be performed without accompanying experience. Nonetheless, we regularly see scientific publications proposing a function for consciousness. A recent Oxford University Press blog post, for instance, claimed that "the function of consciousness is to generate possibly counterfactual representations of an event or a situation," which "hint at the origins of consciousness in the course of evolution" (Kanai 2020).

If one reads it attentively, however, one realizes that the author defined what is meant by "function of consciousness" in a rather counterintuitive manner that contradicts the way any casual reader would interpret the words:

> When we consider functions of consciousness, they are the functions that are enabled by stimuli that enter consciousness or the functions that can be performed only in awake humans or animals. Functions in this sense should not be confused with the question of what kind of effects conscious experiences (or qualia) exert on physical systems. (Ibid.)

In other words, what the author calls the "functions of consciousness" aren't the cognitive tasks *performed by* consciousness, but simply those *visible to* consciousness—i.e. reportable through conscious introspection. Why call these tasks the "functions of consciousness" if they aren't what consciousness *does*, but merely what it 'sees'? According to the author's counterintuitive definition, phenomenal consciousness expressly *isn't* the causative agency behind these tasks—for the author explicitly excludes the causal efficacy of qualia from the definition—but merely their *audience*. As such, the author's theory is entirely beside the point when it comes to the survival value of having qualia or the evolutionary origins of phenomenal consciousness proper.

The impossibility of attributing functional, causative efficacy to qualia constitutes a fundamental internal contradiction in the 'scientific' materialist worldview. There are two main reasons why this contradiction has been tolerated thus far: first, there seems to be a surprising lack of understanding, *amongst materialists*, of what materialism actually entails and implies (see Chapters 2 and 3 of this book). Second, deceptive word games—such as that discussed above—seem to perpetuate the illusion that we have plausible hypotheses for the ostensive survival function of consciousness.

Phenomenal consciousness cannot have evolved. It can only have been there from the beginning, as an intrinsic, irreducible fact of nature. The faster we come to terms with this fact, the faster our understanding of consciousness will progress.

Chapter 5

Consciousness a Mere Accident?

A response to Jerry Coyne

(The original version of this essay was published on my blog, Metaphysical Speculations, *on 14 February 2020)*

Biologist Jerry Coyne has criticized (2020) my argument—discussed in Chapter 4 of this book—that, under the premises of 'scientific' materialism, phenomenal consciousness cannot have been the result of Darwinian evolution. The gist of my argument is that, according to materialism, only *quantitative* parameters such as mass, charge, momentum, etc., figure in our models of the world—think of the mathematical equations underlying all physics—which, in turn, are putatively causally-closed. Therefore, the qualities of experience cannot perform any function whatsoever. And properties that perform no function cannot have been favored by natural selection.

Coyne offers a number of alleged refutations of my claims. He starts by arguing that the qualitative, subjective experiences that accompany the cognitive data processing taking place in our brain may have been merely "byproducts ('spandrels') of other traits that were selected," or even "'neutral' traits that came to predominate by random genetic drift" (Coyne 2020).

Let us take stock of what he is saying here. To begin with, he is implicitly but unambiguously acknowledging my point that consciousness, under materialism, doesn't perform any function. Then, he argues that consciousness could have evolved as a by-product ("spandrel") of the complexity of the brain or even be a merely accidental feature.

The idea of spandrels in evolutionary biology is a contentious one. Many biologists and philosophers criticize it, including Coyne's much-admired Daniel Dennett (1995, 1996). Ian Kluge (n.a.), in his review of Sam Harris's book *Free Will,* also pointed out that

> We could, of course, argue that consciousness and the sense of free will are biological spandrels, i.e. accidental by-products of other evolutionary developments in our brains. One of the problems with this response is that the whole subject of 'spandrels' is bogged down in a definitional debate, i.e. it is not entirely clear what is a spandrel and what isn't. Worse, all examples of spandrels ... do actually serve a function, i.e. they are necessary to achieve something—but that necessity is exactly what epiphenomenalism denies.

Be that as it may, let us charitably ignore this and grant to Coyne that evolutionary spandrels can and do occur. The question then is: Is it at all plausible that phenomenal consciousness is one such a spandrel?

I don't think it is. I can imagine that some relatively trivial, low-cost (in terms of metabolism) biological structures or functions could be merely accidental, but the brain's wondrous putative ability to *produce* the qualities of experience out of unconscious matter is anything but trivial. Indeed, it is nothing short of *fantastic,* the most stunning claim of 'scientific' materialism, the second most important unsolved problem in science according to *Science* magazine (Miller 2005); and now it is a *by-product*?

Materialists have no idea—not even in principle—how the material brain could possibly produce experience. Therefore, they appeal to—and hide behind—the inscrutable complexity of the brain with promissory notes. Phenomenal consciousness—they argue—is somehow an emergent epiphenomenon of that unfathomable complexity, which we one day shall understand. But if such is the case, it becomes unreasonable to posit that something requiring such a level of complexity could have been just an accidental by-product of something else. One can't have it both ways.

At this point, Coyne would probably argue that the brain needed to become complex *anyway,* because natural selection favored higher cognitive ability. And so consciousness just 'came along' for the ride. But we have no reason to believe that the complexity required for more effective cognitive data processing would be the same kind of complexity necessary for the putative emergence of phenomenal consciousness. After all, the complexity underlying better cognitive data processing is meant for ... well, *better cognitive data processing,* not consciousness.

Data processing and experiential states are, in principle, entirely different, even incommensurable domains. On one extreme of the complexity scale, we know that the most powerful computers do not need to have experiential states to perform their functions. On the other extreme, I can imagine a bacterium having experiential states, even though bacteria are some of the simplest living organisms. As a matter of fact, Coyne himself makes this point:

> Any sensation in animals, be they bacteria or humans, involves some sort of qualia. For example, what does it 'feel like' to the crustacean Daphnia to detect a predatory fish in its pond? (2020)

Ironically, Coyne fails to see that these words flirt with some form of panpsychism or idealism: consciousness is already there even in the simplest unicellular organisms; *it doesn't even require a nervous system.* Such a far-reaching and surprising confession contradicts the mainstream materialist storyline—namely, that consciousness is a product or epiphenomenon of (complex) nervous systems—which Coyne believes to be defending. If "any sensation in animals, be they bacteria or humans, involves some sort of qualia," then consciousness is *not* a result of the evolved complexity of the brain, for it doesn't require one. Given this, it is unclear to me exactly where Coyne stands on these crucial issues; his argument doesn't seem to follow any consistent line of reasoning. Is he even really a materialist? Does he understand what materialism is?

Be that as it may, to say that such a fantastic thing as the emergence of phenomenal consciousness from unconscious matter could be a mere spandrel is tantamount to making evolution *unfalsifiable*: if even the most inexplicable of all functions attributed to matter—the one thing that has eluded all attempts at elucidation despite decades of research and speculation—can evolve whether it is at all useful or not, then *anything could have evolved*. We might as well throw our arms up and give up on evolutionary theory altogether, for it would allow us to make no discriminations or predictions whatsoever.

Next, Coyne denies, very emphatically,

> that materialism requires all entities to be measurable. Here's a question: do you have a liver? The answer is based not on measurement, but on observation. I have never heard a definition of 'materialism' that requires quantitative measurement. (2020)

This is a rather embarrassing passage, for it betrays Coyne's startling lack of grasp of the most basic philosophical issues in contention. Here he is alluding to the non-polemical understanding that, under 'scientific' materialism, everything can be exhaustively characterized with *quantities*, such as mass, charge, momentum, etc. What he then claims is that, because *qualities* clearly exist—yes indeed, Jerry, I agree—then materialism *must* allow for the existence of qualities too, not just measurable quantities. Therefore, my claim that materialism attempts to reduce everything to purely quantitative terms can only be wrong.

I wish I were making this stuff up but, alas, I am not.

The point, of course, is not that we can't observe our liver without weighing it or placing a tape measure on it; the point is that, according to materialism, the liver, *in and of itself,* is not *constituted by* the qualities we experience on the screen of perception when we look at it. Instead, it is ostensibly constituted by particles *exhaustively* defined in quantitative terms. It is only when we internally *represent* the liver on the screen of perception that the brain supposedly conjures up, within the boundaries of the skull, the qualities we associate with the liver. It's not a secret, and not even polemical, that this is what mainstream materialism entails. What is surprising is that Coyne—a sworn knight of materialism, of all people—is clearly confused about it.

Indeed, Coyne's argument illustrates precisely the point I made in Chapters 2 and 3 of this book, the original versions of which I had published *before* Coyne wrote his criticism: what he subscribes to is not materialism, but his own idiosyncratic and incoherent misunderstanding of it. This would be forgivable for a casual reader who is not concerned with metaphysics, but not for a man who obviously considers himself a serious participant in the debate. Indeed, that a very vocal and aggressive militant materialist manages to misunderstand what is literally the first thing about materialism—namely, that all qualities are supposedly epiphenomenal—is rather disgraceful.

Piling irony on top of irony, Coyne goes on to quote a passage from the 'Materialism' entry in the *Stanford Encyclopedia of Philosophy*, which he somehow thinks refutes my "definition of materialism." He highlights this segment:

> Of course, physicalists don't deny that the world might contain many items that at first glance don't seem physical—items of a biological, or psychological, or moral, or social nature. But they insist nevertheless that at the end of the day

> such items are either physical or supervene on the physical.

For some reason, Coyne believes that this defeats my point. Unsurprisingly, however, it merely confirms it: according to materialism, the liver may "at first glance [not] seem physical ... but ... at the end of the day [it is] either physical or supervene[s] on the physical." And what is "the physical" under materialism? It is entities exhaustively defined by quantities—such as mass, charge, momentum, geometric relationships, etc.—not qualities; the latter are supposedly epiphenomenal. Therefore, "at the end of the day" the liver, too, is constituted by purely quantitative—not qualitative—physical entities, just as I originally claimed and the rest of the philosophy community knows since freshman year. It is embarrassing that I find myself in the position of having to explain to a militant materialist what materialism is.

Coyne goes on to cite Patricia Churchland, an eliminativist who claims precisely that certain qualities we believe to experience *do not exist at all*, the very opposite of the stance—bizarrely held by Coyne—that materialism also entails qualities. The internal contradictions of his reasoning are just overwhelming. Indeed, next he claims that "we already have lots of evidence that consciousness and qualia are in fact phenomena requiring a materialistic brain, and that manipulating that brain can change or efface consciousness" (Coyne 2020). These assertions instantiate the classical fallacies of question-begging and taking correlation for causation. Allow me to elaborate.

What seems to be beyond Coyne's ability to comprehend is that the dualism between mind and matter he implicitly relies on—particularly when talking about the *mental* effects of *physically* "manipulating that brain"—doesn't exist. To an idealist like me, there is no brain or matter outside or independent of mind. Instead, the 'material' brain is merely the *extrinsic appearance*, in some mind, of the inner mentation of (some other) mind.

When a neurosurgeon manipulates one's brain leading to a

corresponding modulation of inner experience, or when a drug does the same thing after being ingested, what is happening is that a *transpersonal mental process*—whose extrinsic appearance is the surgeon's probe or the ingested pill—*modulates a personal mental process*; namely, the subject's inner experience. This is no more surprising than a thought modulating an emotion, or vice versa. To an idealist, *there is only mind*, matter being just what certain mental processes *look like* from a given vantage point.

As such, the causation link from matter to mind that Coyne relies on to defend his peculiar misunderstanding of materialism is only valid within his peculiar misunderstanding of non-materialist metaphysics. Coyne's views seem to be entirely based on misunderstandings of just about every salient issue.

I believe, thus, that Jerry Coyne just isn't a serious participant in any discussion regarding the nature of mind and reality. As Edward Feser put it, what he writes on philosophy and religion tends to be an "omnibus of fallacies" (2016). Indeed, Coyne's clumsy attempts to defend materialism are a disservice to materialism, a metaphysics that—although fatally flawed—certainly deserves less confused, amateurish treatment.

Chapter 6

Brain Image Extraction

Is it metaphysically significant?

(The original version of this essay was published on my blog, Metaphysical Speculations, *on 14 February 2020)*

Brain image extraction technology has been around for years now: researchers measure a subject's brain activity patterns by means of ordinary electroencephalography (EEG) and are then able to infer the visual experience of the subject during the measurement. This way, they can 'read your mind' or 'extract images' from your brain, so to speak; they can make inferences about your private, first-person visual experience based purely on EEG measurements.

A recent Russian study on brain image extraction (Rashkov *et al.* 2019) has rekindled interest in the subject and may again—understandably, but nonetheless regrettably—lead lay people to the following conjecture: if we are able to translate brain activity measurements into the visual imagery the subject is actually experiencing from a first-person perspective, doesn't that mean that we know how the brain produces experience? Philosophers maintain that we cannot deduce the qualities of experience from purely quantitative measurements, but if—as shown in the Russian study—technology can translate EEG data into visual imagery, surely we have solved the 'hard problem of consciousness' (Chalmers 2003), haven't we?

Surely we haven't. The conjecture—understandable and forgivable as it may be—is totally wrong; it is based on a deep misunderstanding of what is going on here. Allow me to try to explain.

The first thing the Russian researchers did was to take EEG readings of a subject's brain activity while the subject was looking at known images displayed on a screen. In other words, the researchers knew, by construction, what the subject was visually experiencing and what EEG readings corresponded to that experience. They then tuned the internal parameters of a computer algorithm—fancifully called an 'artificial neural network' (ANN)—to capture this known correspondence: that is, to map each EEG reading onto the appropriate image data. This parameter-tuning process is fancifully called 'training.' Notice that no understanding of how the brain putatively produces experience is involved here, since the whole procedure is based merely on *cataloguing empirical correlations* between brain activity patterns and visual imagery.

Let me try to make this clearer with an analogy: you don't need any understanding of how a TV set generates programs in order to determine the empirical correlations between TV channels and programs. Instead, you can simply catalog—by looking at the TV screen while switching channels—which program is being aired on which channel. Moreover, the TV isn't even truly *generating* any program: it is simply receiving and displaying a signal originating from a broadcast station. By the same token, the Russian researchers needed no understanding whatsoever of how the brain putatively produces experience in order to carry out their research. As a matter of fact, their results are even agnostic of whether the brain truly *generates* visual experience at all. All that was required was an empirical correlation between EEG readings and image data, just as there is an empirical correlation between TV channels and programs without the TV having to *generate* the programs.

In practice, the training of the ANN consisted in feeding it an EEG measurement as input and then tuning its internal parameters until it generated—as output—the *same image* the subject was actually looking at when the EEG measurement was taken. We say that this image was the 'target output' of the ANN's training.

Training is performed for many pairs of EEG measurement plus corresponding target output. It goes something like this: imagine that the input is just a number—say, 5—and the target output another number—say, 21. What you then want is to tune the internal parameters of the ANN such that, when it is given 5 as the input, it produces 21 at the output. The result of this tuning could be, for instance, to implement the function $f(\text{input}) = 4 x \text{ input} + 1$, so that $f(5) = 4 x 5 + 1 = 21$, as targeted. Training the ANN consists in finding this function $f(\text{input})$ through trial and error, so the ANN performs an *ad hoc* mapping between input EEG measurements and target output images.

In the case of the Russian study, instead of a single number as input, the ANN receives an array of numbers corresponding to each EEG measurement. Instead of a single number as target output, the ANN receives an array of numbers corresponding to the target images. And then, instead of just one pair of input/target output, it receives several training pairs—that is, a series of EEG measurements, each with its corresponding target image—so the function $f(\text{input})$ generalizes for a variety of inputs. Yet, the essence of what happens during training is what I've just described: the ANN simply implements an *ad hoc* numerical mapping between EEG data and target images, which requires no understanding whatsoever of how or why seeing those images correlates with the given EEG measurements.

That the ANN manages to do this is thus no miracle; it is in fact trivial, the straightforward result of having been trained to do so with image data. The ANN doesn't magically deduce visual qualities from electrochemical patterns of brain activity; it doesn't bridge the explanatory gap between brain function and the qualities of experience; it doesn't even know that its target outputs are images or in any way related to experiences; it just operates on numbers. *Only the researchers—plus you and me—*

know that those numbers correspond to experiences.

The next step in the Russian study was to present the ANN with new EEG measurements that it had not yet seen during training. The idea was to check if the mapping implemented by the ANN was robust enough to remain valid for new data. If so, the output produced by the ANN should be similar to the new images the subject was actually being shown when the new EEG measurements were taken; which was indeed the case. Yet, all this means is that a previously defined numerical mapping between two sets of data was reliable; that's all. None of it has anything at all to do with the hard problem of consciousness or the question of how the brain putatively produces experience.

By explaining how this whole thing works, I hope to have helped you see that neither the Russian study, nor brain image extraction in general, have *any* metaphysical significance. All they establish is that there are reliable *correlations* between patterns of brain activity and inner experience, which was already known. That the science media sometimes portrays these studies as advancing our understanding of how consciousness is putatively produced by the brain is merely a reflection of misunderstanding or deliberate, gullible sensationalism; a bias built into our culture that seeks to find confirmation for the reigning materialist paradigm even at the expense of accuracy and reason.

Chapter 7

The Strange Psychology of Nonsense

How high intelligence manufactures plausibility for 'scientific' materialism

(The original version of this essay was published, under a different title, on IAI News *on 4 March 2020)*

Throughout history, our predecessors in science and philosophy have been convinced that their particular understanding of reality was at least largely correct. Yet, time and again, subsequent generations have proven—or at least were convinced of having proven—them wrong. Each generation has looked upon the ideas of their predecessors as naïve, simplistic, even superstitious.

For instance, during the Renaissance scientists attempted to explain electrostatic attraction as the putative effect of an invisible elastic substance—called 'effluvium'—that supposedly stretched out across bodies. Strange as it may sound now, at the time effluvium was as plausible an explanation for empirical observations as subatomic particles—which are equally invisible beyond the effects they putatively produce—today.

As the Renaissance gave way to the Enlightenment, scientists began trying to frame every phenomenon in terms of the action of small corpuscles—atoms—interacting with each other through direct contact. Any explanation that failed to conform to this template was considered an appeal to magic and, therefore, implausible to say the least. This is why the ideas of an English scientist called Isaac Newton were ignored and even ridiculed for decades: Newton dared to propose that objects attracted one another from a distance by virtue of an invisible, mysterious force he called 'gravity.' We know how that story developed.

As Thomas Kuhn perspicaciously observed in his seminal book, *The Structure of Scientific Revolutions* (2012), changes in science's and philosophy's sense of plausibility aren't monotonic: they don't progress steadily forward, but instead oscillate. Indeed, since Einstein's general theory of relativity, we are back to rejecting the magical action at a distance that Newton thought gravity to be. Now, we have the much more plausible, reasonable, hard-nosed understanding that apples fall to the ground because the Earth ... well, bends the invisible fabric of spacetime around us, as described by an entirely abstract Riemannian geometry that would have made Euclid scoff. Okay.

Please notice that I am not questioning the correctness of our scientific *predictions of nature's behavior,* insofar as they are empirically verified. General relativity unquestionably makes accurate predictions. So did Newton's gravity and—yes—even effluvium in their own time. What I am pointing out is that the way we think about these predicted behaviors—that is, our visualization or mental picture of what is going on—can be regarded as either eminently plausible or utterly implausible depending on the particular historical junction and culture. Our current mental picture of gravity under general relativity—namely, curvatures of spacetime—may be considered utterly implausible in the future, even though that won't change the fact that general relativity makes correct predictions as far as the sensitivity of our measurement instruments today allow us to determine.

In this context, what Kuhn realized was that the mental picture our predecessors in science and philosophy had about what was going on was "produced by the same sorts of methods and held for the same sorts of reasons that now lead to scientific knowledge." Yet, subsequent generations had excellent, even decisive reasons to reject those mental pictures. The inevitable implication is that we structurally believe in nonsense. There is no reason to think that things are different today: future generations are bound to look back at our mental picture of the world and laugh at our myopia and obtuseness, our gullible tendency to appeal to magic.

As a matter of fact, they will have a field day mocking us. For instance, physics is indulging today in the most farfetched feast of appeals to magic ever concocted by the human mind: countless imaginary parallel universes being born each time we glance at the world; space being an illusion somehow derived from time; contradicting views about the origin and early evolution of the universe; the accommodation of complete unknowns by mere labeling, such as the notions of dark matter and dark energy; the list goes on. Effluvium looks very reasonable and benign in comparison.

But what takes the cake isn't the wild speculations of modern physics (a few of them may even turn out to be correct); instead, it is the metaphysics of materialism, which has come to dominate our culture and even our language. Just think of things that matter, things that are immaterial and therefore don't matter, etc. The very root of the word 'matter'—*mater*—means mother, matrix, that from which we came into being.

'Scientific' materialism imagines a purely abstract matrix—namely, matter—that allegedly exists outside and independent of mind, and then tries to explain mind in terms of this abstraction *of* mind. That it then resoundingly fails to catch its own tail doesn't seem to be reason for embarrassment or even weaken materialism's good standing in our intellectual establishment. An obvious appeal to magic as it is—namely, the magic of conjuring up the qualities of experience from the quantities that exhaustively define material arrangements—it is still considered eminently plausible today, just as fairy sorcery once was.

Be that as it may, the key question here is this: Just how is it that we repeatedly end up attributing plausibility to nonsense? What makes us blind to the ultimate untenability of our mental pictures? Why do we regard—with a properly snobbish attitude for good measure—our ludicrous appeals to magic as legitimate, rational, rigorous and even hard-nosed?

I have a little theory about it. At any given point in history, scientists and philosophers always inherit a certain set of foundational values and beliefs—Kuhn famously called it a 'paradigm' (2012)—from the culture they live in. This inheritance defines their sense of plausibility, which is thus also inherited: whatever is validated by their cultural context is bound to sound plausible to them, at least until they examine it more critically. If you and I had grown up with talk of fairies, we would find it entirely plausible that certain odd happenings—such as things being misplaced, disappearing or some people falling mysteriously ill—are caused by fairy sorcery. That we've never seen a fairy wouldn't make them any less plausible than elementary subatomic particles, quantum fields and superstrings: all these invisible entities are imagined purely on account of their alleged *effects*.

The point is that our sense of plausibility isn't at all objective or reliable. What I described above, for instance, is a kind of 'plausibility by habit,' which is almost entirely subjective. In fact, such plausibility by habit is—at least in my view—precisely what keeps materialism alive, despite its unsurmountable problems and internal contradictions.

But then, with time, scientists and philosophers eventually start noticing that their reigning mental picture of reality—I shall call it 'picture 1'—either cannot account for some phenomena or requires modifications and extensions that start to sound implausible even under the values of the reigning paradigm (think of the layers and layers of epicycles in Ptolemaic astronomy, for example). This is the point where a fundamentally new mental picture of reality—'picture 2'—is finally proposed, which tends to focus more or less blindly on addressing the known weaknesses of the previous one. And here lies the problem.

You see, the key psychological motivation for developing

picture 2 is to solve or circumvent the known problems of picture 1. If picture 2 is successful at this task, it tends to be enthusiastically embraced like a longed-for messiah. But the myopia induced by the enthusiasm prevents picture 2 from being critically evaluated *as a whole*, given the complete body of evidence it is supposed to explain. Nobody has interest in kicking all the tires, because everybody is busy celebrating the great advancement that has descended upon us. And so nobody quite sees the *new problems and gaps* that picture 2 introduces.

By the time a new generation of scientists and philosophers starts noticing these problems and gaps, it is too late: a whole new sense of plausibility is now in force; our whole psychology has shifted. The culture is now committed to picture 2 as an advancement, and it naturally doesn't want to give up on this perceived progress. Whatever problems are left must be addressed by incremental additions or adjustments to picture 2, not a new mental picture. We want to believe that we've finally got things right and just miss some details; we want to bank and secure the perceived advance, issuing promissory notes to keep the remaining problems at bay. This is, in fact, exactly what materialists do when confronted with the so-called 'hard problem of consciousness' (Chalmers 2003): "We can't solve it now," they say, "but one day soon we will." And so we keep on waiting.

The advent of 'scientific' materialism during the Enlightenment did solve some problems. The largely religious mental picture of the world that preceded it couldn't account for the regularities of nature's behavior (that is, its seemingly unbreakable 'laws' and automatisms) or the overwhelming suffering and injustice inherent to being alive. As psychiatrist Carl Jung once put it, before materialism we tried to account for far too much in terms of spirit. Hence, a compensatory reaction in the form of metaphysical materialism was to be expected. Moreover, materialism did help the fledgling science of the time to separate its objects of study from the inquiring subject, thereby attaining a level of objectivity that has been instrumental.

Therefore, with great enthusiasm and irresistible momentum, the Western intellectual establishment has embraced materialism and banished the old metaphysics as a relic of superstitious times. But how many scientists and philosophers of the time stopped to notice that materialism, in fact, created more problems than it solved? Who realized, back then, that materialism fundamentally can't explain experience itself, which is all we ultimately know and have? Who, in the 19th century, realized the contradictions that the combined views of metaphysical materialism and Darwinian evolution incurred?

Today we think—by mere force of habit and inherited cultural momentum—that materialism is plausible, even though there is an important sense in which it can't explain *anything* without an appeal to magic. Indeed, materialism is an appeal to one or another magic trick, which we call 'strong emergence,' 'eliminativism' or 'illusionism,' (see Part II of this book) depending on personal taste. Just about everything else is more plausible, if one assesses our metaphysical situation truly impartially.

Materialism has survived thus far because of another kind of trick: in order to defend it and secure our perceived progress since the Enlightenment, intelligent scientists and philosophers—who have staked their public persona and self-image on the validity of materialism—have been deploying their brainpower to *manufacture* plausibility for it.

If an intelligent person is committed to a certain mental picture of the world because of strong—though typically unexamined—psychological investment, it is extraordinary how much they can do to obfuscate the implausibility of the picture, and then manufacture plausibility for it based on a mixture of conceptual conflation, hand-waving and promissory notes. One can basically make anything sound plausible if given enough time and peer support. The history of science and philosophy illustrates this in abundance, but I prefer to provide contemporary examples that are closer to us.

Example one: biologist Jerry Coyne has been so creative at conjuring up plausibility for the notion that consciousness is an evolved trait that he ended up rendering Neo-Darwinism effectively *unfalsifiable* (see Chapter 5 of this book). If you believe Coyne, anything at all could have evolved, irrespective of natural selection. Had a non-materialist argued anything remotely similar, they would have been instantly labeled as irrational.

Example two: because materialism cannot explain experience, some materialists have gone as far as to deny that experience exists in the first place (see Part II of this book). The attempt is to legitimize a kind of insanity for the sake of manufacturing plausibility. And because it is intelligent people who do this, they are able to weave fantastically ambiguous and obscure arguments around their claim; so ambiguous and obscure that it becomes effectively impossible to figure out what they are *actually* saying, if anything. For instance, if you ask a consciousness-denier whether they deny—yes or no—the pain they would presumably feel if they were tortured, they would say they don't. But if you then point out that their answer implies that consciousness actually exists, they would say, "No, what we call pain is merely a *functional state*, not actual experience." "But then you do deny our felt pain!" you might add, just to watch—in understandable frustration—the conversation loop back to the beginning and repeat itself. Philosopher Galen Strawson called this peculiar maneuver 'looking-glassing' (2013), but I prefer to call it 'plausibility by obfuscation.'

Example three: to manufacture plausibility for the current paradigm *as a whole*, intelligent scientists and philosophers are—ironically—prepared to sacrifice the plausibility of *any one element* of the paradigm. For instance, experimental results in quantum physics have now refuted physical realism: there is no physically objective, standalone world of tables and chairs out there (see Chapters 16, 17, 20 and 21 of this book). The only way to avoid this empirical conclusion is to postulate a mindboggling number of new, undetectable, parallel but *real* physical universes being magically created every time someone or something merely looks at the world. Or else we have to accept that the physical world of tables and chairs exists only insofar as it is observed. Which option do you think is less *im*plausible? Renowned physicist Sean Carroll is convinced it is the former (Chen 2019). And he is not embarrassed to admit it, for we live in a culture in which his preference for magic is—remarkably—not regarded as ludicrous. The price Carroll is willing to pay to manufacture plausibility for materialism is to turn physics itself into a caricature (Hossenfelder 2020).

Example four: as I discussed more extensively in Chapter 2 of this book, the main reason why materialism is considered plausible by the average educated person is that they don't understand what materialism means. In other words, what people think of as plausible isn't actual materialism, but a misunderstanding that passes for materialism. Specifically, many educated people fail to understand that, according to mainstream materialism, the colors they see, the tones they hear, the textures they feel, the flavors they taste and the aromas they smell exist *only inside their skull*. The world of their experiences is supposedly entirely within their head. What is out there is devoid of all qualities of experience; it can't even be visualized, for visualization always entails qualities. The best you can do is to picture it as a purely abstract realm of ghostly silhouettes and mathematical equations; but even this goes too far. Indeed, famed mainstream physicist Max Tegmark has posited that the world out there is, literally, pure mathematics (2014). Now, how many casual materialists would still consider materialism plausible if they truly grokked this? Remarkably, ignorance is a great tool to help manufacture plausibility for materialism.

And so, just like every generation before us, we enthusiastically continue to embrace nonsense; we enthusiastically manufacture plausibility to preserve mistakes with which we've become identified.

The only way to break out of this cycle is to realize that mind excels at deceiving itself. As the history of science and philosophy attests to, this is what mind does best. And the very idea that we can be objective investigators impartially assessing the world around us is mind's greatest self-deception. Our mental pictures of the world are inherently unreliable. The day we have the courage to bite this bullet at a cultural level, is the day we will begin to make real progress.

"But what kind of progress?!" I hear you ask indignantly. For if we can't trust our own mind, how can we progress? Remarkably, this act of seeming intellectual suicide opens up a solid avenue of understanding; one utterly unlike the conceptual quicksand in which we have so far found ourselves stuck.

By understanding and acknowledging that mind—if given the chance—always deceives itself, we lose almost everything. *But not quite everything.* One final option remains viable: *all there is is mind.* Reality is a *mental construct*; it consists of mind tricking itself into believing that there is something outside mind. Because the collective madness of our mainstream worldview has thoroughly infected our language, I can't even begin to convey to you how obvious this seeming absurdity is. Instead, you will think *I* am crazy. And that's fine too.

Just beware: to say that everything is in mind doesn't mean that everything is in *your or my* mind alone. For even the notion that you and I have our own private mind—separate from others—is part of the self-deception. To truly understand what this hypothesis means, one has to dig deep, very deep into layer upon layer of ingrained and unexamined assumptions inherited from culture and by now almost hardwired into our DNA. Most people can't or won't go there; the mere attempt exposes them to what I like to call the 'vertigo of eternity': the appalling realization that what is *actually* going on is not even commensurable with what they *think* is going on.

Language—at least our present language—cannot do justice to reality; the latter escapes conceptualization. But through scrupulous conceptual reasoning we can still get close to it, which is precisely what I have been attempting to do for over a decade with my body of work. To progress we must see through the self-deception; we must grok how the plausibility-manufacturing industry creates a hall of mirrors around us; we must realize that those who scream "reason!" the loudest are often the most deluded and unreasonable ones. And then, one day, there might just be a generation who will look back to their forefathers and finally say: "Darn, they did get it right."

Part II

On Consciousness Denialism

Chapter 8

The Mysterious Disappearance of Consciousness

What makes otherwise intelligent, highly educated people deny the undeniable?

(The original version of this essay was published on IAI News *on 9 January 2020)*

Phenomenal consciousness is regarded as one of the top unsolved problems in science (Miller 2005). Nothing we can—or, arguably, even could—observe about the arrangement of atoms constituting the brain allows us to deduce what it feels like to smell an orange, fall in love or have a bellyache. Remarkably, the intractability of the problem has led some to even claim that consciousness doesn't exist at all: Daniel Dennett (1991) and his followers famously argue that it is an illusion, whereas neuroscientist Michael Graziano proclaims that "consciousness doesn't happen. It is a mistaken construct" (2016). Really?

The denial of phenomenal consciousness is called—depending on its particular formulation—'eliminativism' or 'illusionism.' Its sheer absurdity has recently been chronicled by Galen Strawson (2018), David Bentley Hart (2017) and yours truly (Kastrup 2015: 59-70), so I won't repeat that argumentation here. My interest now is different: I want to *understand* what makes the consciousness of an otherwise intelligent human being deny its own existence with a straight face. For I find this denial extremely puzzling for both philosophical and psychological reasons.

Don't get me wrong, the motivation behind the denial is obvious enough: it is to tackle a vexing problem by magically wishing it out of existence. As a matter of fact, the 'whoa-factor' of this magic gets eliminativists and illusionists a lot of media attention. But still, what kind of conscious inner dialogue do these people engage in so as to convince themselves that they have no conscious inner dialogue? Short of assuming that they are insane, fantastically stupid or deliberately dishonest—none of which is plausible—we have an authentic and rather baffling mystery in our hands.

The only way to go about elucidating the mystery is to investigate, with patience and an open mind, the arguments offered by eliminativists and illusionists. The cover story of a recent issue of *New Scientist*, for instance, sensationally announced the discovery of the "True nature of consciousness: Solving the biggest mystery of your mind" based on an essay by Michael Graziano (2019). In it, Graziano argues—predictably—that consciousness doesn't actually exist.

He starts the essay by defining his usage of the term 'consciousness': "it isn't just the stuff in your head. It is the *subjective experience* of some of that stuff" (*Ibid.*, emphasis added). Clearly, thus, Graziano is talking about *phenomenal* consciousness, not the other technical usages of the term (Block 1995). Phenomenal consciousness entails the *subjective experiences* that seem to accompany the material stuff going on in your head. So Graziano's challenge is to persuade you that, despite all appearances to the contrary, those experiences don't actually exist.

His argument rests on the idea that consciousness is adaptive: it is undoubtedly beneficial to us to recognize and understand ourselves as agents in our environment—i.e. to have a model of ourselves—if we are to survive. In this context, Graziano argues that consciousness is merely a model the brain constructs of itself, so it can "monitor and control itself" (*Ibid.*). Consciousness seems immaterial—his argument goes—simply because, in order to focus attention on survival-relevant tasks, the model fails to incorporate superfluous details of brain anatomy and physiology. In Graziano's words, "the brain describes a simplified version of itself, then reports this as a ghostly, non-physical essence" (*Ibid.*).

This is all very reasonable. The problem is that *it has nothing to do with phenomenal consciousness*. Graziano's authoritative prose disguises a sleight of hand: he implicitly changes the meaning he attributes to the term 'consciousness' as he develops his argument. He starts by talking about subjective experience—i.e. *phenomenal* consciousness, which is what science can't explain—just to end up explaining something else entirely: our ability to cognize ourselves as agents and metacognitively represent our own mental contents. If anything, Graziano's argument *presupposes* phenomenal consciousness: once raw experience is assumed to be in place, then—and only then—does his argument help us understand how such experience can be configured so as to enable reflective introspection and a felt sense of self.

What Graziano describes as a "ghostly" or "ethereal essence"—and then proceeds to explain away in terms of brain function—is merely a colloquial understanding of consciousness, one that regards it as something akin to a 'soul.' This, of course, isn't the technical issue in contention; it isn't what is meant by phenomenal consciousness. What it feels like to lift a heavy bag, have your tongue burned by hot tea or hit your head against a wall isn't "ethereal" at all (try the wall if you doubt me). There is remarkably little in Graziano's argument to justify the rather ambitious title of his essay.

Also recently, Keith Frankish—an illusionist—published an essay on *Aeon* making the case that consciousness is, well, an illusion (2019). Never mind the fact that illusions are inherently *experiential* and therefore *presuppose consciousness*; the subtitle of his essay—"Phenomenal consciousness is a fiction written by our brains"—gave me hope that he would at least face the core issue head-on, instead of throwing a smokescreen of conceptual obfuscation and hand-waving.

Disappointingly, however, Frankish already starts out by conflating science with the metaphysics of materialism and then weaving a blatantly circular argument:

> It is phenomenal consciousness that I believe is illusory. For science finds nothing qualitative in our brains, any more than in the world outside. The atoms in your brain aren't coloured and they don't compose a colourful inner image. (*Ibid.*)

The argument structure here is the following:

1. Material things, in themselves, have no qualitative properties (like color, flavor, etc.), only our perceptions of them do;
2. The brain is a material thing;
3. From (1) and (2), the brain has no qualitative properties;
4. Experience is reducible to the brain;
5. From (3) and (4), experience cannot entail qualitative properties.

Ergo, phenomenal consciousness cannot exist; it must, instead, be an illusion—or so his argument goes.

Notice, however, that step (4) blatantly begs the question: it *presupposes* materialism, which is precisely the metaphysical point in contention. Ironically, what Frankish actually accomplishes is to highlight an implication of materialism that reduces it to absurdity.

His next point—the core of his case—doesn't fare much better. He explains:

> it is useful to us to have an overview or 'edited digest' (Dennett's phrase) of [our brain] processes—a sense of the overall shape of our complex, dynamic interaction with the

> world. When we speak of what our experiences are like, we are referring to this sense, this edited digest. (*Ibid.*)

His point is that, when we introspect, what we experience isn't our brain processes as they are in themselves, but an inaccurate, distorted, "edited digest" thereof. This is the basis of Frankish's claim that experiences are illusions: they are *mis*portrayals of what they represent, *mis*representations of material brain states. That's why—the argument goes—a bellyache feels nothing like networks of firing neurons inside our head, even though the latter is supposedly what the ache actually is.

Misportrayals as they may be, since Frankish's basic premise is that only material states exist, these 'edited digests' must themselves consist of material brain states as well—what else? And thus, infinite regress is upon us: since the brain states corresponding to the misportrayals feel nothing like networks of firing neurons, they must themselves be misrepresented by some meta-introspective system. But alas, the resulting meta-misportrayals also necessarily consist of material brain states, so we need a meta-meta-introspective system that misportrays the misportrayals of the misportrayals; and so on. No amount of material indirection can make material states seem experiential, just as no number of extra speakers can make a stereo seem like a television and no number of extra legs can make a centipede fly: the two domains are just incommensurable. All Frankish accomplishes is to conceptually postpone the inevitable confrontation with the actual problem at hand.

To his credit, Frankish does explicitly address the obvious objection against illusionism: that the qualitative properties of experience—color, flavor, etc.—cannot be illusions, for illusions themselves entail experiential properties. Here is the passage wherein he tries to tackle this objection (if you fail to understand his prose, no worries, I summarize it below rather simply):

> Properties of experiences themselves cannot be illusory in the sense described, but they can be illusory in a very similar one. When illusionists say that phenomenal properties are illusory, they mean that we have introspective representations like those that we would have if our experiences had phenomenal properties. And we can have such representations even if our experiences don't have phenomenal properties. Of course, this assumes that the representations themselves don't have phenomenal properties. But, as I noted, representations needn't possess the properties they represent. (Ibid.)

What he is saying is that, whether we have actual experiences—phenomenal properties—or not, everything can happen *as if* we had them. However, this succumbs to the exact same objection it was meant to rebut: for things to happen as if we had experiences, it must *seem* to us that we do have experiences, even if we don't. But Good Lord, the 'seeming' is already an experience in and of itself. The introspective representations must themselves be experiential, otherwise there would be no 'seeming,' no illusion. Frankish is tying himself up in tortuous conceptual knots in his clumsy attempt to perform an impossible magic trick; to abstract the concreteness of experience away.

Bewilderingly to me, he then makes a remarkable admission: "But how does a brain state represent a phenomenal property? This is a tough question" (*Ibid.*). Well, this is the *only* salient question, isn't it? On the answer to this question rests Frankish's entire case. He continues:

> I think the answer should focus on the state's effects. A brain state represents a certain property if it causes thoughts and reactions that would be appropriate if the property were present. (*Ibid.*)

This blatantly begs the question again. Only under the

assumptions of eliminativism or illusionism do effects sufficiently account for the question Frankish is leaving open. What defines experience is precisely that, *regardless of its effects,* there is something *it is like to have it.*

While acknowledging that he faces an explanatory challenge, Frankish suggests that all metaphysics face the same challenge:

> it is not only illusionists who must address this problem. The notion of mental representation is a central one in modern cognitive science, and explaining how the brain represents things is a task on which all sides are engaged. (*Ibid.*)

I regard this as outright misdirection. Yes, the mechanisms of mental representation in general aren't fully understood, but this is not the salient issue here. What is salient is this: *only illusionists have to account for the experience of 'seeming'*—i.e. illusion—*while denying experience to begin with.*

It is, of course, conceivable that I've failed to properly grasp what Frankish and Graziano are trying to say. But if someone with my background can't understand the arguments they make in non-academic publications meant for the general public, I don't think the burden is on me to make the next move in the debate. The mind-bogglingly extraordinary claim that phenomenal consciousness—the carrier of all our knowledge, the one thing we can be absolutely sure of—doesn't exist requires rather extraordinary substantiation. Otherwise, it is legitimate to conclude that eliminativism and illusionism are precisely what they seem to be: blatant nonsense.

I believe Frankish's and Graziano's arguments are based on question-begging, conceptual obfuscation and sleights of hand. They betray, in my opinion, remarkable lack of lucidity, clarity of thought and ability to introspect into one's own experiences so as to cognize what most of us mean by the word 'consciousness.' Where does this leave us regarding the mystery I originally set out to elucidate?

My present opinion is that illusionists and eliminativists are sincere; but also so fanatically committed to a particular metaphysics—materialism—that they inadvertently conjure up, and then tie themselves in, perplexing webs of conceptual indirection, ultimately deceiving themselves. In their inner dialogue, I suspect they implicitly replace the obvious meaning of the term 'consciousness' with one or another secret conceptual abstraction, and then strive towards proving that such abstraction doesn't actually exist. Well, guess what? Of course it doesn't! They defer tackling the salient questions with layer upon layer of smoke and mirrors just to admit, at the very end, that the questions haven't actually been addressed. However, by adding and then wrestling with all those artificial in-between layers, they get the impression that progress has been made, only one step being left at the end. But in fact *nothing* has been accomplished; nothing at all. The 'problems' they solve aren't real problems to begin with, just conjured-up artifacts of conceptual fog. There is nothing of any substance or relevance prior to the "tough question" of "how does a brain state represent a phenomenal property" (*Ibid.*) if experience—as they allege—doesn't exist.

Despite all this, here we are, discussing eliminativism and illusionism as if they were serious hypotheses because—bewilderingly—they have somehow acquired a degree of academic respectability. Such is the sorry, surreal state in which we find our Western philosophy: it has managed to find *the* most implausible and self-defeating proposition *conceivable* to the human mind, and now tries to defend it with a straight face. What would Parmenides and Plato have thought of it? More gravely, what will future generations think of *us*?

Chapter 9

The Mysterious Reappearance of Consciousness

A Rejoinder to Michael Graziano

(The original version of this essay was published on IAI News *on 23 January 2020)*

Princeton neuroscientist Michael Graziano has replied (2020) to the original publication of the essay in Chapter 8 of this book, wherein I've criticized the bizarre notion—publicly defended by Graziano and called 'eliminativism' or 'illusionism' in philosophy—that phenomenal consciousness, with its felt qualities, doesn't actually exist. The present essay is a rejoinder to Graziano's reply.

As attentive readers will have immediately noticed, the most conspicuous thing about Graziano's reply is his undisguised, unabashed *failure to actually address my criticism*. Indeed, he ignores my core charge that what he putatively refutes isn't phenomenal consciousness at all—despite his insistency that it is—but merely a certain mistaken sense of self. I interpret such a failure as an implicit admission that my charge was right on the mark for, otherwise, it's hard to imagine that Graziano would have missed the chance to issue a corrective of some kind. Instead, he chose to pursue a trite and puerile series of *ad hominem* attacks and straw men—whereby he labels me a mystic New Ager—in a rather blatant attempt to deflect attention from the real issues in contention.

Let us look at all this in more detail. Graziano begins by suggesting that the myriad philosophical positions being debated today about the nature of consciousness can be divided into only two camps: mysticism and materialism. As a philosopher of mind familiar with those debates, I couldn't help but smile at the remarkably simplistic naiveté of such a suggestion. For Graziano, if you are not a materialist, you are a mystic.

He goes on to suggest that my criticisms come from a "nonscientific, or often pseudoscientific, political side" and reflect the "wooly thinking of philosophy that's lost its integrity" (2020). Let us ignore the strange allusion to politics in what is—or at least should be—an eminently technical debate; Graziano seems to conflate science and philosophy: either my argument is philosophical or (pseudo)scientific. I'm afraid it can't be both.

Indeed, Graziano's defining claim is overtly philosophical: "consciousness doesn't happen. It's a mistaken construct" (2016). The closest scientific claim would be to say that certain *contents of* consciousness do not *correspond to* objective facts, but that's not what he asserts. Confusingly, he then proceeds to directly contradict his own claim:

> Among the most common and puzzling reaction I get goes [*sic*] something like this: "Graziano says that consciousness does not exist; that we lack an inner dialogue; that getting stuck by a pin, or walking into a wall, is ethereal." None of these statements are true (Ibid., emphasis added)

Well, if it's not true that "Graziano says that consciousness does not exist," then what is it that he denies? How are we to reconcile this reply with his defining claim that "consciousness doesn't happen" (Graziano 2016)?

Illusionists and eliminativists play a slippery and deceptive game of words that philosopher Galen Strawson once called 'looking-glassing' (2013): they implicitly and conveniently change the meaning they attribute to the word 'consciousness' depending on circumstances. When they make the fantastic, headline-grabbing claim to have circumvented the hard problem of consciousness (Chalmers 2003), they can only mean phenomenal consciousness by it—that is, felt experience, such as the felt pain of a pin prick. But when their argument is shown to actually *presuppose* felt experience—as opposed to denying it—they claim to mean *something else* by the word 'consciousness.' Graziano's reply is loaded with looking-glassing: attempts to have the cake and eat it too.

So what exactly is the 'consciousness' that Graziano is now, in his reply to me, denying? He explains (I quote relatively extensively to avoid misrepresenting his case):

> The truth is that the chair you think is there is not exactly the same as the chair that is actually there ... Information about the chair enters your visual system; your brain builds a simplified, 'quick-and-dirty' version (a simulation or model as it's sometimes called) [of it]; your cognition has access to that model; as a result, you can talk about the chair. ... [But] the models are simplifications; they are not perfectly detailed or accurate. ... You claim to have a conscious experience. You make that claim because you think it's true – your higher cognition has hold of that information. ... But that information is almost certainly not perfectly accurate. Therefore, we know – I would say with certainty – that whatever consciousness you actually have, it is different from the consciousness that you think you have. (2020)

In other words, our brain builds internal representations—"models"—of the world. What we actually experience in consciousness are these internal representations. But the representations are not identical with what is actually out there in the world. Instead, they are simplifications thereof. Therefore, what we experience in consciousness doesn't correspond accurately with what is objectively out there.

This is all perfectly reasonable and fine. In fact, it is even trite. But now I ask you: Does any of this deny consciousness? Do the inaccuracies of our internal representations deny the consciousness where these inaccuracies are *experienced*? By acknowledging that our simplified representations indeed are experienced, is Graziano denying or in fact *presupposing* phenomenal consciousness?

Graziano's argument could certainly justify the following assertion: some, most, perhaps even all contents of consciousness—i.e. felt experiences—do not accurately represent objective facts. I have no problem accepting this, not least because it's trivially true and trite. But the assertion certainly does not imply that consciousness itself "doesn't happen. Is a mistaken construct" (Graziano 2016). It is justifiable to say that some *contents of* consciousness are mistaken constructs in the sense of being simplified models of the world or even the self, but not that the felt experience of said mistaken constructs doesn't happen. To me this is so obvious I blush to have to state it.

But Graziano then proceeds to construct a questionable logical bridge:

> Some of the attributes of consciousness that you claim to have, you probably don't have. Some of the attributes of consciousness that you actually have, you probably don't know that you have. (2020)

It doesn't follow from (a) the fact that some of our internal representations are inaccurate that (b) we mistake what consciousness itself is. The bridge between the two claims is a *non sequitur*. Nonetheless, I accept—independently of Graziano's argument—that some of what we believe to experience isn't actually experienced. For instance, most people think they experience their entire visual field in high resolution, while in fact this is the case for only a tiny area in the middle of the field. The illusion of high resolution arises from our continuous, subliminal scanning of the world by rapidly moving our eyes.

However, even the illusion presupposes consciousness: to think that we experience our entire visual field in high resolution is itself an experienced thought. Even if all our internal representations were totally wrong, they would still be experienced as such. After all, a completely illusory life is precisely a life that unfolds *only in consciousness*, not one that denies it. To say that consciousness isn't what we think it is neither entails nor implies that consciousness doesn't exist; on the contrary: it presupposes the consciousness that is deluded about what characteristics it attributes to itself. Is any of this hard to understand?

Interestingly, Graziano explicitly acknowledges, in his reply, the existence of felt experience, phenomenal consciousness:

> An internal dialogue? Sure, of course we all have it. A mind spinning with thoughts and sensory impressions? Yes. The pain of being stuck by a pin, happiness, memory, a moment of decisiveness, a moment of indecision? Yes, all of that is present, in some form. *Consciousness? Yes, indeed.* (2020, emphasis added)

He thus confirms precisely what I stated in my original criticism: his alleged refutation isn't of consciousness in the phenomenal sense. What is it, then, that he denies? It is precisely what I originally claimed:

> when we introspect, when we dip into our intuitions and thinking, we report something totally different – not electrical impulses and synapses, not interacting chunks of information, but something amorphous and *ghost-like*. Philosophers have been trying to put words to that *fuzzy extra*

> *essence* for millennia, but none of the vocabulary really pins it down. (2020, emphasis added)

Hence, exactly as I claimed in my original essay, what Graziano denies is the "amorphous, ghost-like essence" we colloquially associate with the idea of a 'soul,' which has little to do with phenomenal consciousness. So much for a rebuttal of my original criticism. Ironically, Graziano makes my case arguably better than I did.

So why all the fuss? The problem is that (a) making the case that some of our internal representations are inaccurate, (b) claiming that we don't actually have some of the experiences we consciously think we have, or (c) asserting that an ethereal, ghost-like sense of self is illusory are all relatively trivial and trite arguments; none of them would make big headlines, for none address the hard problem of consciousness. The 'easy' problems Graziano is trying to tackle have little to do with the challenge of making sense of phenomenal consciousness. If he had never attempted to portray his work otherwise, I would probably have never heard of him and we would not be having this discussion right now.

Graziano concludes his peculiar 'reply' by taking what he seems to believe is a shrewd jab at me:

> I can understand the visceral dislike, maybe even fear, from people who think this scientific approach encroaches on their sense of mystery. (2020)

Maybe I should indeed be terrified. My only hope is that, at the end of the day, reason and clarity of thought will hold more sway than misleading word games, conceptual confusion and hand-waving.

Chapter 10

No Ghost, Just a Shell

A rejoinder to Keith Frankish

(The original version of this essay was published on my blog, Metaphysical Speculations, *on 10 April 2020)*

Like neuroscientist Michael Graziano, illusionist philosopher Keith Frankish has also replied (2020) to my original criticism of illusionism and eliminativism, which is reproduced in revised form in Chapter 8 of this book. This essay is a rejoinder to Frankish's reply.

10.1 An initial reflection

In criticizing illusionists such as Frankish, one is always faced with the dilemma of either writing with the general public in mind or the individual illusionist one is criticizing. The most effective line of reasoning is different in each case, for the public isn't tied up in the conceptual and definitional knots illusionists create for themselves. Indeed, whereas the public—watching from a more objective, uncommitted vantage point—can easily grasp the blatant circularity and inconsistency of the illusionist argument, the illusionists themselves are too immersed in their own story to fathom any of it. Instead, one must first meet them where they are, otherwise they will choose to believe that their points are merely misunderstood by their critics.

Having always written with the general public in mind, it thus comes as no surprise to me that Frankish should feel certain that I do not grasp what he is saying. At no point in his reply does he seem to entertain the possibility that I actually understand perfectly well where he is coming from, why he thinks what he thinks, and yet still consider his story blatantly absurd.

As a matter of fact, in the early years of my career as a computer engineer, I wrestled intensely with the question of how to build computers that would consider themselves conscious even if not programmed to do so; that is, how to construct a machine that would not only perform calculations, but also spontaneously claim to experience these calculations, just as you and I claim to experience the goings-on in our brain. This wasn't armchair philosophizing for me, but a very concrete and practical question. And that's precisely why I ended up wasting so much time on it: I never stopped to examine the implicit assumptions embedded in the very problem statement that motivated my search.

And so it was that, in the first years of the 21st century, Pentti Haikonen, a researcher at Nokia, came up with a computer architecture that would not only consider itself conscious, but—Haikonen thought—also in fact be conscious (2003). Haikonen's deeply insightful realization was two-fold: first, the original semantic anchoring of the input signals fed into the computer should be preserved—as opposed to being encoded into arbitrary binary symbols—if the computer is to consider itself conscious; second, feedback loops should be inserted in the architecture at strategic points, so as to allow the computer to introspect by re-representing its own computational activity.

Haikonen's approach, which I recognized as brilliant and hold in very high esteem to this day, can be regarded as effectively elaborating—much more specifically and persuasively than Frankish himself—on Frankish's claim that introspective (mis) representation is what leads to the belief that we are conscious. As such, and implausible as it may sound to him, I believe I actually understand why Frankish considers illusionism so compelling. I am very familiar with the thinking and motivations behind it, in a fairly high level of (engineering) detail.

Indeed, Haikonen creatively tackled many of the difficulties I had identified for building a machine that could spontaneously claim to be conscious. Alas, we would be unable to verify such claim, for the only way to know would be to *be* the machine. Yet, the claim alone would already be a remarkable engineering achievement, one I was very interested in contributing to.

When it comes to us, however, it's not just a matter of making spontaneous claims: we actually *know* that we are conscious, for we *are* ourselves. In our case, therefore, we must address the hard problem of consciousness (Chalmers 2003), which Haikonen's architecture—despite his philosophically naïve claims to the contrary—completely fails to do: instead of creating consciousness in the phenomenal sense, his approach merely *presupposes* it. And so does Frankish's.

No amount of structure, complexity, feedback, recursion, re-representation, etc., can make a fundamentally unconscious substrate produce experience, in the same way that no amount of added speakers can turn a stereo into a television, and that no amount of extra legs can make a centipede fly. Recursive re-representations can only *complexify* preexisting experiential states, not create them from something fundamentally non-experiential. More specifically, what recursive re-representations can do is to make preexisting phenomenality accessible to metacognitive introspection, not create it.

10.2 Conflating consciousness with metacognitive awareness

For Frankish, it is our ability to introspect by metacognitively re-representing our neural processes that characterizes what we call 'consciousness.' He writes:

> It is a mark of conscious experience that we are, or can easily become, aware of having it. We can direct our attention inward ('introspect') and think about the experiences we are having. (2020)

It is this introspection that, according to Frankish, creates the illusion of qualitative experience: "Our introspective systems monitor these [neural] processes but misrepresent them as a simple quality," he says. "The illusion concerns the nature of these processes—the belief that they are simple qualia." Later he continues: "It is this emphasis on the effects of introspection that makes the notion of illusion so appropriate here" (*Ibid.*).

The problem is that Frankish conflates *phenomenal* consciousness—that is, raw experience, 'what-it-is-likeness'—with *meta*-consciousness. As Jonathan Schooler explained,

> Periodically attention is directed towards explicitly assessing the contents of experience. The resulting meta-consciousness involves an explicit re-representation of consciousness in which one interprets, describes, or otherwise characterizes the state of one's mind. (2002)

But phenomenal consciousness does not require meta-consciousness: if an experience falls outside the field of our attention, we have the experience without being aware *that* we are having it. For instance, we regularly experience our breathing without metacognitive representation. Moreover, as discussed by Jennifer Windt and Thomas Metzinger, dreams largely lack introspective re-representation, despite their undeniably experiential nature (2007). Even the emerging 'no-report paradigm' in neuroscience (Vandenbroucke *et al.* 2014, Tsuchiya *et al.* 2015) rests on the understanding that experience *can*—and frequently *does*—occur without explicit introspective awareness, such as in the cases of blindsight that Frankish likes to cite.

In conflating consciousness with meta-consciousness, Frankish is failing to heed a key conceptual distinction seminally discussed by philosopher Ned Block (1995). Phenomenal consciousness, in turn, doesn't need to be introspectively accessible in order to exist. These are two different things.

Therefore, Frankish's appeal to introspective (mis) representation to explain experience away is based—as I originally claimed—on conceptual confusion: if experience isn't there to begin with, we have no reason whatsoever to believe that introspective re-representations would be, or even seem to be, experiential either. Instead, everything would happen 'in the dark,' without the light of awareness.

10.3 Eating the cake and having it too

For Frankish's position to have any relevance in helping tackle or circumvent the hard problem of consciousness (Chalmers 2003), what he must deny is *phenomenality, felt experience, qualia, 'what-it-is-likeness.'* Anything else, despite potentially having some other philosophical application, would be irrelevant as far as the hard problem is concerned.

Unsurprisingly, thus, Frankish often emphasizes that what he denies is precisely phenomenality, qualia, experience. For instance, already in the subtitle of a recent essay, he wrote that, "Phenomenal consciousness is a *fiction* written by our brains" (Frankish 2019, emphasis added). This doesn't seem to leave much room for ambiguity, as philosophers use the qualifier 'phenomenal' precisely to specify, unambiguously, that what is meant by the word 'consciousness' is qualities, felt experiences, 'what-it-is-likeness.'

But to deny the qualities of experience is to deny, for instance, that we feel pain; is to say that our agonizing screams under torture—as well as the accompanying physiological processes—are merely functional, useful for getting help; but that, from the inside, none of the dreadful qualities we associate with pain are actually felt. If Frankish denies the qualities of experience, presumably he wouldn't mind undergoing torture, which I very much doubt to be the case.

And so, in his reply to me, Frankish already starts out by

acknowledging that it is "utterly ridiculous" to claim that "people are not conscious, don't have experiences." He goes on to say that "illusionists don't deny that we are conscious," that there is a "sense in which we undoubtedly are conscious," that "our lives are filled with conscious experiences" (2020). Really? What is it, then, that illusionists *do* deny?

"What illusionists reject is a certain conception of what consciousness is," Frankish claims (*Ibid.*). But then again, if illusionism is to have any relevance as far as the hard problem is concerned, the "conception of consciousness" that must be denied is precisely that entailing pain and emotion, felt experiences, which Frankish has just acknowledged to exist! Any other conception of consciousness—such as Block's 'access consciousness' (1995) or Schooler's 'meta-consciousness' (2002)—is *irrelevant* for the hard problem: it still leaves us with having to explain how raw experience, whether metacognitively represented or not, arises from an allegedly non-experiential substrate.

Frankish thus faces an impossible dilemma, which he can only tackle by systematically contradicting himself. He deserves our sympathy, for the job of manufacturing even a smidgen of plausibility for what is the most incongruous maneuver conceivable to the human mind—that of denying itself—is not exactly easy. On the one hand, he must acknowledge that "illusionists don't deny that we are conscious" (Frankish 2020), otherwise they would just be crazy. On the other hand, he also has to claim that

> Illusionists reject the qualitative conception of consciousness. They hold that qualia, and the private show they constitute, are illusory; they seem to exist but don't really. This is the core claim. (*Ibid.*)

How are we to square this circle? Frankish seems to be making a distinction—which he conspicuously doesn't elaborate upon—between experience or phenomenality on the one hand, and felt qualities on the other. He acknowledges the former while, bizarrely, denying the latter. Yet, to reject the "qualitative conception of consciousness" (*Ibid.*) is to deny experience, phenomenality, phenomenal consciousness itself; after all, the latter is *defined* as entailing the felt qualities of experience.

Is Frankish playing some silly game of words? In what sense is he acknowledging that we have pain and emotion if he is denying the qualities that pain and emotion *are*? Granted, pain and emotion are associated with certain functions and behaviors, but this has nothing to do with consciousness. By focusing on function and behavior to the exclusion of qualia, Frankish is merely *ignoring* the hard problem, closing his eyes to it, not tackling or circumventing it in any meaningful sense.

When he reassures us that "illusionists don't deny that we are conscious" (*Ibid.*) and thus aren't outright crazy, Frankish is appealing to our intuitive understanding of conscious states as felt qualities. But then, having accomplished that, he immediately turns around and rejects the "qualitative conception of consciousness" (*Ibid.*) so as to portray his approach as relevant to addressing the hard problem, instead of something utterly trivial. Which one is it? He can't have it both ways. Either he is sane, or his work is relevant when it comes to the hard problem.

10.4 Explanation by redefinition of terms

Contrary to what Frankish suggests, the qualitative dimension of experience isn't a merely conceptual reality, but a *felt* and *immediate* one. It is very important that we keep this in mind.

You see, there are many entities in science whose only accessible reality is conceptual: think of imaginary numbers in mathematics or quantum fields in physics, for instance. The world behaves *as though* these conceptual entities existed and, as such, it is very useful to imagine that they do. But we have no immediate, felt access to them; all we know about them is our conception of them.

Therefore, these conceptual entities are perfectly amenable to being redefined, if doing so helps to make sense of things. For instance, it has been useful to redefine gravity as a curvature of spacetime, instead of an invisible force acting between two bodies from a distance. We have no direct acquaintance either with the curvature of spacetime or the invisible force, so we might as well feel free to redefine gravity based on theoretical convenience.

However, an analogous rationale does *not* apply to phenomenal consciousness, for the qualities of experience aren't merely conceptual; they are *immediately felt*. By rejecting "a certain conception of what consciousness is" (Frankish 2020) illusionists aren't making these felt qualities disappear; they are merely ignoring them, pretending that they don't exist.

Indeed, whatever definition of consciousness we choose to use in our conceptual games, there remains this thing—this undeniable thing immediately accessible to us prior to all conceptual reasoning—that will continue to exist whatever we call it. The hard problem of consciousness is essentially about this thing, not the word 'consciousness.' If you think the label 'consciousness' shouldn't be applied to it, fine, I don't care, call it something else; call it ... well, the 'thing.' But the thing won't cease to exist just because you renamed it. Even if we can't appropriately define it in words, it won't be affected; it will remain what it is and has always been. Terminology games don't change reality, no matter how hard we wish they did.

Frankish, however, seems to think that he can make the felt qualities of experience—the thing—disappear simply by redefining terms. Consider the following passages from his reply:

> "experiences are physical states of the brain"
>
> "consciousness consists, not in awareness of private mental qualities, but in a certain relation to the public world"
>
> "It is this global broadcasting and its effects that constitute consciousness"
>
> "I am proposing that consciousness is this complex of informational and reactive processes"

Hey, I can play this game too. How about "consciousness is the involuntary microscopic twitching of my left big toe"? Or—just a little more seriously—"consciousness is the collapse of the quantum wave function in the synaptic clefts of my prefrontal cortex"? In terms of explanatory power, are these merely definitional statements really so different from "consciousness is a complex of informational and reactive processes"? Are they any less arbitrary? Do definitional statements have any explanatory power at all? Do they solve any problem?

The bottom-line is this: we know first-hand what consciousness is, regardless of how the word is defined. It doesn't matter how often and how passionately Frankish repeats his statements of faith, it is this thing we know that matters; it won't disappear because of semantic games. Substituting redefinitions of terms for actual argument just won't do. Otherwise, I would have won the *Fields Medal* long ago by merely redefining yet-unsolved problems in such a way that the solution would be trivial.

10.5 Failure to grasp the criticism

Frankish's central point is that our introspective re-representations of our own physical brain states are illusory in the sense that they don't *accurately* portray said brain states. That's why—in his view—we mistakenly think we have qualia, instead of just physical brain states: the latter are metacognitively misrepresented as seeming qualia, which is the putative illusion in question.

My original refutation of this argument was as simple as it was generic: if the misrepresentations *seem* qualitative, then the very *seeming* is already a quality, regardless of what exactly the misrepresentations seem like. After all, an illusion is already a *felt experience in and of itself*, regardless of its lack of representational accuracy. The implication is that we *do* have qualia, not despite our re-representations being inaccurate in the way Frankish claims them to be, but precisely *because* of it.

Yet, Frankish failed to understand this simple point of mine. He misconstrues and misportrays it as something unnecessarily more restrictive, which can be seen in the following passage (if you find it too difficult to follow his reasoning, no worries, I summarize it in simpler words below):

> Could we seem to have qualia without really having them? Kastrup thinks not. "Good Lord," he exclaims, "the 'seeming' is already an experience in and of itself." Does this simple point blow illusionism out of the water, as Kastrup supposes? There's one way it might. Suppose that ... seeming to perceive a thing involves being aware of the mental qualities one would have been aware of if one were really perceiving it. Then, by analogy, seeming to introspect a mental quality would involve being aware of the mental quality one would have been aware of if one were really introspecting it. And that, presumably, is the very same mental quality. The illusion would involve a real instance of the thing that was supposed to be illusory! ... The flaw in this objection is obvious: *it assumes that experience involves awareness of mental qualities*. (Frankish 2020, emphasis added)

What Frankish is saying here is that my criticism holds if, *but only if*, our alleged misrepresentations of perceptual brain states correspond to "the mental qualities one would have been aware of if one were really [consciously] perceiving" (*Ibid.*). But such constraint is not at all necessary for my criticism to hold: whether the alleged misrepresentations match what would have been the actual qualia of perception or not is irrelevant, as long as the misrepresentations *seem like something; anything*; it doesn't matter what. The seeming alone already entails felt qualities—whatever they may be—and, therefore, felt qualities must exist.

Even if what is misrepresented are physical brain states—as opposed to experiential qualities—the corresponding seeming, in and of itself, is already an experience entailing its own (illusory) felt qualities. Therefore, contrary to what Frankish claims, the only assumption my criticism makes is that *there is seeming*, which is precisely what illusionism requires (otherwise one cannot speak of illusions to begin with).

10.6 Final reflections

For many years I have earned my living doing corporate strategy in the high-tech industry, perhaps the most rewarding but also most unforgiving environment for analytic thinking. In that world, even subtle and understandable failures of reasoning are very quickly—and often disproportionately—punished, either by management or by the markets. Reality can always be counted upon to settle all questions in a rather brutal but objective manner; something I have grown to appreciate over the years, for it forces me to be constantly critical of my own narratives.

It strikes me, though, that in philosophy one seems to be able to get away with incoherent thinking indefinitely. If one cannot clearly and substantively argue for one's own position, verbose misdirection, ambiguity and handwaving often sustain just enough doubt about whether one is actually wrong. And so, nonsense can survive *ad infinitum*. There is always the lingering doubt that, hidden behind impenetrably obscure language constructs and indecipherable conceptual acrobatics, there might just be some deep, non-obvious philosophical insight. Yet, often there is none; often things are precisely what they seem to be: very confused and self-contradictory thinking. Perhaps this is the reason why philosophy doesn't seem to make progress.

Although I have made a deliberate effort in this essay to patiently meet the illusionists where they are—so as to do something different than just repeating my original criticisms, most of which Frankish didn't even address, such as my claim that he falls for the fallacy of infinite regress—it remains my position that illusionism is the most ludicrous and self-defeating view conceivable. Nothing in the history of human thought is, or can be, more preposterous than it. That some otherwise intelligent people espouse it is, in my view, merely a psychological artifact—a desperate attempt to salvage an untenable metaphysics many have associated their very identity with—not the outcome of clear, rational thought.

My willingness to engage in an extensive, detailed and protracted debate about illusionism should, therefore, not be construed as a sign of respect for it; I have precisely none. In fact, I find it embarrassing to be in a position of having to argue against it. My doing so reflects merely a begrudging acknowledgment that our philosophy is in a lamentable state and that, if anything is to be done about it, it must be done from *within* our present circumstances.

In a less confused world, illusionism wouldn't even be a joke. Perhaps we will get there one day. In the meantime, however, brutally honest, even scathing public criticism may be the only system of checks-and-balances available to preserve the sanity of philosophy. The present essay has been written in this ultimately well-meaning spirit.

Part III

On Constitutive Panpsychism

Chapter 11

Panpsychism Ultimately Implies Universal Consciousness

There is no plausible halfway compromise between materialism and idealism

(The original version of this essay was published, under a different title, on IAI News *on 8 January 2020)*

There is an arguably bizarre theory in philosophy today—gaining momentum in both academia and popular culture—called 'panpsychism.' It has many variants, but the most recognizable one posits that elementary subatomic particles—quarks, leptons, bosons—are conscious subjects in their own right. In other words, the idea is that there is something it feels like to *be* an electron, or a quark, or a Higgs boson; their experiential states are allegedly an irreducible property of the particles themselves, just like mass, charge or spin. According to this theory—which has been openly embraced by influential mainstream figures, including reductionist neuroscientist Christof Koch (2012a)—our complex conscious inner life is constituted by an unfathomable combination of the experiential states of myriad particles forming our brain.

I fully understand the urge to circumvent the failures of mainstream materialism, according to which matter is all there truly is, experience being somehow an emergent epiphenomenon of certain ephemeral material arrangements. There is growing awareness in both science and philosophy that materialism is untenable, as I discussed in Chapter 1 of this book. The question is whether simply *adding*—next to mass, charge, spin—fundamental experiential properties to matter is a persuasive and legitimate way out, or just evades the need for explanation.

You see, I can easily accept that my cats are conscious, perhaps even the bacteria in my toilet. But I have a hard time imagining—especially when I am eating—that a grain of salt contains a whole community of little conscious subjects. The panpsychist's motivation for wanting even the humble electron to be conscious is to treat experiential states in a way analogous to how physical properties are treated in chemistry: as the physical properties of particles combine in atoms, molecules and aggregates to give rise to emergent macroscopic properties—such as the wetness of water—the panpsychist wants the experiential states of particles in our brain to combine and give rise to our integrated conscious inner life. The idea is to fold experience into the existing framework of scientific reduction and emergence, therein residing most of the appeal and force of panpsychism.

To do so, the panpsychist takes subatomic particles to be discrete little bodies with defined spatial boundaries. This way, their respective experiential states are thought to be encompassed by such boundaries, just as our human experiences seem to be encompassed by our skull. Indeed, since each person's consciousness does not float out into the world, but is personal in the sense that it is limited by the boundaries of the person's body, so subatomic particles are imagined by the panpsychist as discrete little bodies, each containing separate and independent subjectivities.

The panpsychist then posits that the inherent subjectivity of different particles can combine into compound subjects if and when the particles touch, bond or otherwise interact with one another in some undefined chemical manner. Notice that this approach makes sense only through analogy with physical properties: the mass of an electron is 'held' within the electron's boundaries, therefore it's only logical—the argument goes—that its experiential states should also unfold within the same boundaries. Or is it?

The problem is that subatomic particles aren't discrete little bodies with defined spatial boundaries; the latter is a simplistic and outdated image known to be wrong. According to Quantum Field Theory (QFT)—the most successful theory ever devised, in terms of predictive power—elementary particles are just local patterns of excitation or 'vibration' of a spatially unbound quantum field. Each particle is analogous to a ripple on the surface of a lake: we can determine the location of the ripple and characterize it through physical quantities such as the ripple's height, length, breadth, speed and direction of movement. Yet, there is nothing to the ripple but the lake: we can't lift it out of the lake, for the ripple is merely a pattern of movement of the water itself. Analogously, according to QFT, an elementary subatomic particle is just a pattern of excitation or 'vibration' of an underlying quantum field. Like the ripple, we can determine the particle's location and characterize it through physical quantities such as mass, charge, momentum and spin. Yet, there is nothing to the particle but the underlying quantum field. The particle *is* the field, 'moving' in a certain manner.

What is fundamental in nature is the quantum field, not the elementary subatomic particle it happens to form through excitation or 'vibration'; after all, the latter is, by definition, reducible to the former. The panpsychist is thus forced to attribute consciousness not to the particle, *but to the underlying field itself*. The particle represents just a particular modulation or configuration of experience, not the creation of consciousness out of unconsciousness. Panpsychism is physically coherent only if the quantum field is conscious *as a whole*, as a unitary subject. And because the field doesn't have spatial boundaries, panpsychism implies universal consciousness and fails to explain our own personal subjectivities.

Here the panpsychist could counterargue that the physical properties of an elementary subatomic particle—such as mass, charge and spin—are localized and belong to the particle, not to the whole quantum field. After all, the particle's mass, charge and spin are akin, in the analogy above, to the height, length and breadth of the ripple, which are indeed local properties of the ripple, not of the whole lake. Therefore—the argument goes—why can't we say that experiential states, too, belong to the particle alone, not to the quantum field as a whole?

To see why this doesn't work, notice first that one can easily deduce or predict the quantitative parameters that define a ripple—e.g. height, length, breadth—from the equally quantitative parameters that describe the behavior of the lake. Physicists do it all the time in the field of fluid dynamics. Deducing the quantitative physical properties of a particle from the quantitative physical parameters that describe the underlying quantum field is entirely analogous. There is thus no fundamental problem in deducing quantity from quantity.

However, deducing *quality* from quantity is something entirely different. Experiential states are qualities; they cannot be exhaustively described in quantitative terms. No numerical parameter can tell someone with congenital blindness what it feels like to see red; or someone who never fell in love what it feels like to, well, fall in love. Indeed, this is precisely the so-called 'hard problem of consciousness' (Chalmers 2003) that plagues mainstream materialism and motivated the creation of panpsychism in the first place. *One cannot make an unconscious quantum field give rise to a conscious particle for exactly the same reasons that one cannot make an arrangement of matter give rise to experience.* Therefore, once again, the panpsychist either defeats their own purpose or must attribute consciousness to the quantum field *as a whole*, as a fundamental property *of the field*, which implies universal consciousness and fails to explain our private inner lives.

Granted, this is not what the panpsychist was bargaining for. For in light of this insight, experiential states can no longer be treated analogously to physical properties in chemistry.

Experience is no longer local, encapsulated in little bodies of matter—as physical properties can still be imagined to be—but 'smeared out' across spacetime instead. The entire rationale for explaining our conscious inner life through the combination of discrete experiential states at a microscopic level goes out the window: there is nothing to be combined within the boundaries of our skull anymore, just spatially unbound, universal fields and their patterns of excitation. Panpsychism cannot explain private, individual experience.

This is a *coup de grâce* for panpsychism, for the idea that microscopic subjects of experience can somehow combine to form seemingly unitary, macroscopic ones already constitutes a 'hard problem' in its own right (Chalmers 2016a): What kind of magical interaction between two particles could possibly have the extraordinary effect of combining two fundamentally distinct fields of experience? Even the logic underpinning panpsychism is faulty: the panpsychist attributes to the *subject of perception* a structure discernible only in *that which is perceived*. That the physical world we see seems 'pixelated' at the level of elementary subatomic particles may be an artifact *of the screen of perception*, not a reflection of the structure of the perceiver. As an analogy, notice that the image of a person on a computer screen appears pixelated when looked at closely. Yet, that doesn't mean that the person is herself made of discrete rectangular blocks! The pixelation is an artifact *of the screen*, not the structure of the person represented on it. By the same token, that our body is made of subatomic particles says something about how we are represented on the screen of perception, not necessarily about the subject that does the perceiving.

Don't get me wrong: the panpsychist is going in the right direction when they consider consciousness irreducible, and such openness is a valuable commodity in our overwhelmingly materialist culture. My hope is that, freed from the missteps discussed above, the panpsychist finds intellectual space to contemplate a more promising alternative—one that entails leaving every vestige of materialism behind, instead of striking a halfway, Frankenstein-monster-like compromise. The idea is that, *in lieu* of preserving physical properties alongside experiential states as fundamental aspects of nature, the way to go is to *reduce the physical to the experiential.*

You see, every scientific and philosophical explanation entails reducing a phenomenon to some other aspect of nature, different from the phenomenon itself. For instance, we reduce or explain a living organism in terms of organs, organs in terms of tissues, tissues in terms of cells, molecules, atoms and subatomic particles. But because we can't keep on explaining one thing in terms of another forever, at some point we hit rock-bottom. Whatever is then left is considered to be our 'reduction base': a set of fundamental or irreducible aspects of nature that cannot themselves be explained, but in terms of which everything else can. Under materialism, the elementary subatomic particles of the standard model—with their intrinsic physical properties—constitute the reduction base.

To circumvent materialism's failure to explain experience, the panpsychist simply *adds* experience—with all its countless qualities—to the reduction base. Arguably, this is a cop-out: inflated reduction bases don't really explain anything, they just provide subterfuge for evading explanations. A good rule of thumb is that the best theories are those that have the smallest base, and then still manage to explain everything else in terms of it. On this account, panpsychism just isn't a good theory.

Good alternatives to materialism are those that *replace* elementary particles with experiential states in their reduction base, as opposed to simply adding elements to it. We call this class of alternatives 'idealism.' And then the best formulations of idealism are those that have one single element in their reduction base: universal consciousness itself, a spatially unbound field of subjectivity whose particular patterns of excitation give rise to the myriad qualities of empirical experience. Under such a theory, a unified quantum field *is* universal consciousness.

There is nothing absurd about this theory; the common impression that there is, is just a knee-jerk reaction of our current intellectual habits. As a matter of fact, the theory is arguably the most parsimonious, internally consistent and empirically sound view yet devised. Importantly, as I have extensively elaborated upon elsewhere, idealism—unlike panpsychism—can explain how our private, personal subjectivities arise within universal consciousness. I therefore hope that the momentum gathered by panpsychism in both academia and popular culture is transferred, intact, to this uniquely viable avenue of inquiry, before the inherent shortcomings of panpsychism discourage—as they are bound to eventually do—those seeking an alternative to materialism.

Chapter 12

Sentient Robots, Conscious Spoons and Other Cheerful Follies

How blind spots of critical thinking are distorting our collective intuitions of plausibility

(The original version of this essay was published on Scientific American *on 7 March 2018)*

Contemporary science fiction seems obsessed with ideas such as downloading consciousness into silicon chips, sentient robots, conscious software and whatnot. Films such as *Her* and *Ex Machina*, as well as recent episodes of series such as *Black Mirror*, portray these ideas very matter-of-factly, desensitizing contemporary culture to their extraordinary implausibility.

The entertainment media takes its cue from the fact that research on artificial *intelligence*—an objectively measurable property that can unquestionably be engineered—is often conflated with artificial *consciousness*. The problem is that the presence of intelligence does not entail or imply the presence of consciousness: whereas a computer may effectively emulate the information processing that occurs in a human brain, this does not mean that the calculations performed by the computer will be accompanied by private inner experience. After all, the mere *emulation* of a phenomenon *isn't* the phenomenon: I can emulate the physiology of kidney function in all its excruciating molecular details in my desktop computer, but this won't make the computer urinate on my desk. Why, then, should the emulation of human information processing render a computer conscious?

When it comes to consciousness, even academics seem

liable to lose touch with basic notions of plausibility. This is because, despite the prevailing assumption that consciousness is generated by arrangements of matter, we have no idea how to deduce the qualities of experience from physical parameters. There is nothing about mass, charge, spin or momentum that allows us to deduce how it feels to see red, to fall in love or to have a bellyache. Rationally, this abyssal explanatory gap should immediately lead us to question our prevailing assumptions about the nature of consciousness. Unfortunately, it has instead given license to a circus of ungrounded and largely arbitrary speculations about how to engineer, download and upload consciousness.

Already in the early 20th century, Bertrand Russell observed that science says nothing about the intrinsic nature of the physical world, but only about its structure and behavior (2009). A contemporary of Russell's, Sir Arthur Eddington, also observed that the only physical entity we have *intrinsic* access to is our own nervous system, whose nature is clearly experiential (1928). Might this not be the case for the rest of the physical world as well? Under this panpsychist hypothesis, the explanatory gap disappears: consciousness isn't generated by physical arrangements but, instead, is the intrinsic nature of the physical world. The latter, in turn, is merely the extrinsic appearance of conscious inner life.

Many panpsychists, however, go a step beyond this otherwise reasonable inference: they posit that consciousness must have *the same fragmented structure* that matter has. This way, individual subatomic particles are posited to be conscious subjects in their own merit, in that they allegedly have private inner experiences of their own. And because the body of more complex subjects—such as you and me—is made of subatomic particles, our conscious inner life is inferred to be constituted by a combination of the conscious perspectives of countless little subatomic subjects. Some interpretations of panpsychism imply even that inanimate aggregations of matter are conscious: a relatively recent essay claimed, "The idea that everything from spoons to stones is conscious is gaining academic credibility" (Goldhill 2018).

Oh well.

The notion that inanimate objects are subject to their own experience may sound absurd; and it is. However, the reason to dismiss it is not intuition—conditioned as the latter is by unexamined cultural assumptions—but simple logic. You see, the movement from "consciousness is the intrinsic nature of the physical world" to "subatomic particles are conscious" relies on a flawed logical bridge: it attributes to *that which experiences* a structure discernible only *in the experience itself*. Allow me to elaborate.

The concept of subatomic particles is motivated by experiments whose outcomes are accessible to us only in the form of conscious perception. Even when delicate instrumentation is used, the output of this instrumentation is only available to us as perception. Those experiments show that the images on the screen of perception can be divided up into ever-smaller elements, until we reach a limit. At this limit, we find the smallest discernible constituents of the images, which are thus akin to pixels. As such, subatomic particles are the 'pixels' of *experience*, not necessarily of the experienc*er*. The latter does not follow from the former.

Therefore, that living bodies are made of subatomic particles does not necessarily say anything about the structure of the experienc*er*: a body is itself an image on the screen of perception and so will necessarily be 'pixelated' insofar as it is perceived. Such pixelation reflects the idiosyncrasies *of the screen of perception*, not necessarily the structure of the subject itself. As an analogy, the pixelated image of a person on a television screen reflects the idiosyncrasies of the television screen; it does not mean that the person herself is made up of pixels.

I thus submit that consciousness is indeed the intrinsic nature of the physical world, but subatomic particles and other inanimate objects are not conscious subjects in their own merit. After all, as Freya Mathews pointed out, the boundaries of inanimate objects are merely nominal (2011)—where does the river stop and the ocean begin?—whereas those of conscious subjects are unambiguously determined by, for instance, the range of the subjects' internal perceptions. So inanimate objects cannot be conscious subjects.

With inanimate objects excluded, only living organisms and the inanimate universe *as a whole* can be conscious subjects (a more extensive argument for this point can be found in Part IV of this book). This way, as a living nervous system is the extrinsic appearance of an organism's inner experiences, so the inanimate universe *as a whole* is the extrinsic appearance of *universal inner experiences*. Circumstantially, the inanimate universe at its largest scales has indeed been found to structurally resemble a nervous system (Vazza & Feletti 2017). Under this view, there is nothing it feels like to be a spoon or a stone, for the same reason that there is nothing it feels like to be—at least as far as you can assess through introspection—one of your neurons in and of itself. There is only something it feels like to be your nervous system *as a whole*—that is, you. Analogously, there is only something it feels like to be the inanimate universe *as a whole*.

If biology is the extrinsic appearance of conscious subjects other than the inanimate universe itself, then the quest for artificial sentient beings boils down to abiogenesis: the artificial creation of biology from inanimate matter. If this quest succeeds, the result will again be *biology*—artificially created as it may be—not computer emulations thereof. The differences between flipping microelectronic silicon switches and metabolism are hard to overemphasize, so nature gives us no reason to believe that a collection of flipping switches should be what private conscious inner life looks like from the outside; let alone stones and spoons.

Emboldened by empirically ungrounded and illogical notions being noisily contemplated in some corners of academia, the entertainment media is rendering nonsense culturally plausible. As a result, a whole generation is growing up taking folly for future likelihood.

Part IV

On Analytic Idealism

Chapter 13

An Overview of Analytic Idealism

How a consciousness-only metaphysics, 100% science-compatible, solves the problems of materialism and panpsychism

(The original version of this essay was published, under a different title, in the Blog of the American Philosophical Association *on 16 October 2019)*

I defend a modern, analytic formulation of metaphysical idealism—descriptively called 'analytic idealism'—according to which the ground of existence is phenomenal consciousness. Everything else—I maintain—is reducible to configurations and patterns of excitation of consciousness. This does not mean that spoons and home thermostats are conscious in and of themselves; that's panpsychism. Neither does it mean that reality is in your or my individual mind alone; that's solipsism. Instead, I acknowledge that other living beings have a conscious inner life of their own. I also acknowledge that there is something out there, beyond individual minds, which would continue to exist even if no one were looking at it. However, in my view, this 'something out there' is itself *experiential* in nature—that is, it consists of *trans*personal mental activity. Such mental activity merely *presents* itself to us *as* the inanimate universe. From this perspective, analytic idealism is consistent with how David Chalmers defined 'objective idealism' (2018).

At the same time, experimental results emerging from the field of foundations of physics make clear that, whatever is out there, beyond individual mentation, does not have a definite state before it is observed (see Chapters 16 and 17 of this book).

In other words, the external environment, *as it is in itself,* does not comprise objects with definite form, position, momentum, etc. It consists instead of superposed *possibilities* or *tendencies*. Metaphysically, I interpret this as follows: the transpersonal mental activity that surrounds us is best understood as ambiguous *thought processes* of the kind we experience, for instance, when we weigh different possible decisions without being sure of which one to take. Therefore, although there indeed is a world out there, this world isn't physical in the sense we ordinarily attribute to the word; physical properties result, instead, from an *interaction* between our own mental processes and the transpersonal mental processes within which we live. This interaction is what physicists call 'observation' or 'measurement,' which cognitively amplifies one of the superposed possibilities out there, leading to the impression that we inhabit a definite physical world. As such, the physical world is merely an image in the individual mind of the observer; each one of us perceives our own *physical* world, as defined by the context of our own observations. Analytic idealism can thus also be regarded as 'subjective idealism' (Chalmers 2018) with respect to physicality. But don't get me wrong: I do believe we share a common environment independent of us all; it's just that this common environment does not comprise, in itself, the properties we associate with physicality.

All this, of course, immediately raises the following question: What is our relationship, as minded individuals, to the hypothesized transpersonal mind that surrounds us all? As someone who considers parsimony a key guiding value in philosophy, I maintain that there is ultimately *only one, universal consciousness*. I think we, along with all other living beings, are merely dissociated mental complexes—'alters'—of this fundamentally unitary universal mind. This is akin to how a person suffering from dissociative identity disorder also manifests multiple disjoint centers of awareness. The boundary of dissociation is what separates us from our environment and each other. The way this boundary presents itself on the screen of perception is what we call our skin and other sense organs. As experienced from the inside—that is, from a first-person perspective—each living being, plus the inanimate universe as a whole, is a conscious entity. But as experienced from the outside—that is, from a second- or third-person perspective—our respective inner lives present themselves in the form of what we call matter, or physicality. Indeed, in my view 'matter'—*all* matter—is merely the name we give to what conscious inner life *looks like* from across its dissociative boundary. That's why there are such tight correlations between inner experience and measurable patterns of brain activity.

Finally, an important element of analytic idealism is that the transpersonal mental processes that underlie and ground the inanimate universe do not necessarily entail metacognition. This may need some brief unpacking: metacognition is our human ability to explicitly evaluate our own mental activity, which requires more than just raw phenomenal consciousness. An experience is metacognitive if, in addition to having the experience, the subject also knows *that* they have the experience. Metacognition enables deliberation, reasoning and planning. Purely instinctive thought processes, on the other hand, are those that, despite being conscious, lack metacognition. Now, because the laws of nature are seemingly stable and predictable, I maintain that the transpersonal mentation underlying the inanimate universe is instinctive, not metacognitive. After all, instinctive behavior is regular and predictable, just like the laws of nature seem to be. As such, universal consciousness does not necessarily have a plan; it may be doing what it is doing merely because it has the innate disposition to do so.

It's fair to place analytic idealism in the historical context of German idealism, even though I do not endorse the view that universal consciousness is rational and deliberate.

Analytic idealism is very well aligned with the ideas of Arthur Schopenhauer, as discussed in his *magnum opus*, *The World as Will and Representation* (Kastrup 2020). Just as Schopenhauer thought that underlying all nature is a 'blind' will, which presents itself on the screen of perception as matter, I maintain that instinctive mental processes—most likely of a volitional character, as the universe's movements and evolution suggest volitional impetus—underlie physicality. With Schopenhauer, and departing from Immanuel Kant, I maintain that we can make sound inferences about the nature of the *noumena*—i.e. things in themselves—through personal *introspection*; for unless we are prepared to accept an arbitrary discontinuity in nature, if my inner mentation presents itself to outside observation in the form of the matter constituting my nervous system, then the matter of the rest of the universe, too, should be the extrinsic appearance of (universal) conscious inner life. As Schopenhauer put it, "we must learn to understand nature from ourselves, not ourselves from nature."

In the context of contemporary analytic philosophy, analytic idealism can be regarded as a variation of cosmopsychism (Chalmers 2018). It is important to notice, however, that I do not look upon phenomenality as a fundamental aspect or property *of* matter, the latter also having other properties or aspects. No. To me what we call 'matter' is merely the *extrinsic appearance of inner phenomenality*, as observed from across a dissociative boundary. There's nothing more to it. To say the same thing in a different way, 'matter' is the handy label we give to the contents of a particular modality of experience; perception. Therefore, and paraphrasing Bishop Berkeley, as far as matter is concerned 'to be is indeed to be perceived,' even though the mental activity underlying matter continues to exist whether it is observed from the outside or not—that is, whether it is apprehended *as matter* or not.

With Kant and Schopenhauer, I do think the contents of

perception—by which I mean particular arrangements of perceptual qualities such as color, flavor, smell, etc.—are mere *representations* or *phenomena* of the world as it is in itself. What I have been calling 'extrinsic appearances' are thus at least largely equivalent to Schopenhauer's 'representations' and Kant's 'phenomena.' I maintain that these representations or phenomena arise from the *interaction* between our own private, dissociated mentation and the transpersonal mentation constituting the environment that surrounds us; in modern language, they arise from 'observation' or 'measurement.' In positing that representations or phenomena are what conscious inner life looks like from across a dissociative boundary, I am attempting to add to Kant's and Schopenhauer's insights: neither seems to have explained precisely how representations/phenomena arise from noumena/will, respectively.

Still with Kant and Schopenhauer, analytic idealism maintains that spacetime itself is merely an evolved cognitive scaffolding in the perceiver's individual mind, which is then populated by representations. The world as it is in itself—Kant's noumena or Schopenhauer's will—is incommensurable with spacetime and is not constituted by the qualities of perception. Running the risk of excessive anthropomorphization—which Schopenhauer himself guarded against—we could think of it as transpersonal thoughts driven by instinctive volitional urges. These transpersonal thoughts merely *present* themselves to us, at a glance, as a tapestry of color, flavor, smell, etc., because encoding our apprehension of the surrounding environment in this manner had significant evolutionary advantages (Kastrup 2018a).

According to analytic idealism, the spacetime scaffolding and the basic percepts that populate it are *cognitive mechanisms* we evolved as a species, not standalone existents. As such, they are *built into the organism*—not passed on by culture—being no more fluid than the organism itself is. Having said that—and backed by modern psychology—I do think our ordinary experience of the world entails much more than just basic percepts. What we actually experience is, in large part, a narrative we build with, and then project onto, the percepts. In other words, we don't just apprehend raw 'pixels'—meant here metaphorically and generally, so one could speak e.g. of auditory 'pixels'—but partition, group and weave them together according to a story we tell ourselves subliminally. Unlike the basic percepts, this inner narrative is passed on by culture and education. It is precisely one such culture-bound story that leads most people today to look at the world outside and see discrete objects made of matter outside mind. After all, the discreteness of objects is merely nominal (Mathews 2011), whereas the mind-independent status of the world is merely a theoretical abstraction (Kastrup 2018b). This substitution of the concreteness of the world by abstractions would have sounded ridiculously absurd only a few hundred years ago.

The notion that spatially-unbound phenomenal consciousness is the ground of existence is extremely exciting, for it circumvents all the insoluble problems of today's metaphysics. The core of the 'hard problem,' for instance, is this: first, we infer that the world is made of matter outside and independent of consciousness; then we imagine that certain patterns of organization of matter—such as our brain—can, somehow, give rise to consciousness; finally, we infer that the material dynamisms of the external environment modulate the experiences generated by our brain through the mediation of the sense organs. The latter is what we call perception. The problem is that there is nothing about the abstract, quantitative parameters that describe and define material organization in terms of which we could deduce the qualities of experience. I could exhaustively describe the material system we call an 'apple' in terms of its constituent particle masses, momenta, charges, spatiotemporal positions, etc., but none of that would give me any insight into what it feels like to see the redness, or taste the sweetness, of the apple. We fundamentally can't bridge the gap between physical *quantities* out there and experiential *qualities* in here.

Under analytic idealism, however, what is out there is experiences too, even though experiences qualitatively different from those on the screen of perception. In other words, what it feels like to *be* the world out there is qualitatively different from what it feels like to *perceive* such world. But bridging the gap between two different sets of qualities is empirically trivial: we witness it happening all the time. For instance, the qualities of our thoughts can translate directly into the qualities of our emotions: there is something it feels like to have the thought that, say, life has no meaning, which then translates into the felt emotion of hopelessness or despair. The quality of the thought, although different from the quality of the emotion, directly leads to the latter. Therefore, there is nothing difficult about the hypothesis that transpersonal thoughts out there, upon impinging on the dissociative boundary of our respective alter—whose representation is our skin and other sense organs—translate into the qualities of perception. There is no unbridgeable gap anymore.

The equally insoluble 'subject combination problem' (Chalmers 2016a) of constitutive panpsychism is also circumvented by analytic idealism. There is no need to explain how fundamentally disjoint, microscopic subjects of experience—such as those hypothetically corresponding to the subatomic particles that form our brain—combine to constitute the seemingly unitary, macroscopic subject we seem to be. Analytic idealism already starts from a universal subject, so nothing needs to combine. The challenge it must then tackle is precisely the opposite: How does one universal consciousness seemingly divide itself up into multiple individual subjects, such as you and me? While this is a legitimate problem, it is one whose solution nature has already given us in the form of the psychiatric condition called dissociation. Whether we understand the inner mechanisms of the condition or not, we know empirically that mental space can seemingly split itself up into multiple, cognitively disjoint, co-conscious centers of awareness. I insist on 'seemingly' because we also know, empirically, that such split is reversible and merely apparent: patients have been known to overcome dissociation and re-integrate their alters into a unitary, internally-connected mental space. The suggestion here is that universal consciousness can undergo something akin to dissociation, thereby forming multiple disjoint alters, such as you and me. What we call 'life' or 'biology' is the extrinsic appearance, the representation of this dissociation; that is, life is what a dissociative process at a universal level *looks like* when observed *from across its dissociative boundary*. This, in my view, is all there is to life.

Analytic idealism has enormous relevance for our lives both as individuals and members of a collective. After all, our metaphysical views, even if implicit and unexamined, color every aspect of our lives, from our moral values to our sense of meaning. The notion that existence, at its most fundamental level, is sentient and unitary has tremendous implications for how we regard each other, the planet where we live and the universe at large. To mention one obvious example, the philosophical school of existentialism seems to presuppose separation, which analytic idealism fundamentally denies, even though it allows for the *appearance* of separation.

But the most relevant implication of analytic idealism has to do with how it informs our understanding of death. If life is the extrinsic appearance of dissociative processes at a universal level, then death—the end of life—is *the end of dissociation*; that is, the re-integration of our conscious inner life into a broader context. In an important sense, this flips our understanding of death upside-down: death is no longer the end or infinite constriction of consciousness, but precisely an *expansion*. As a matter of fact, there are compelling empirical indications that this is the case, as discussed in Chapter 25 of this book.

Analytic idealism can also inform our understanding of personal identity. As Schopenhauer already explained over 200 years ago, we each have a kind of double identity or "twofold existence," as he put it (cf. Kastrup 2020). The first is what he described as "the eternal world-eye," which "looks out from all knowing creatures." Philosopher Itay Shani described essentially the same thing in modern analytic terms under the label "core-subjectivity," which is the "dative ... of experience [i.e.] that to whom things are given, or disclosed, from a perspective" (2015). Core-subjectivity entails no contents—no name, place of birth, profession, age, episodic memories, etc.—and no narratives of self-identity. Instead, it consists of an empty subjective space with its inherent, undifferentiated feeling of primal 'Iness.' You can imagine it as what it would feel like if you became completely amnesic, but still conscious, while in an ideal sensory deprivation chamber.

Because it's undifferentiated and content-free, core-subjectivity is *identical* in each and all of us: we are all "the one eye of the world which looks out from all knowing creatures." Indeed, it is because of the unbroken continuity of our core-subjectivity over time that we believe we are the same entity since birth, even though everything else about us—our body, thoughts, opinions, self-image, memories, etc.—has changed multiple times over since then.

Framing it in my terms, core-subjectivity is intrinsic to universal consciousness; there may even be an important sense in which it *is* universal consciousness. Therefore, death—the end of dissociation—changes nothing about it. Death happens *within* core-subjectivity, not *to* it. The recipient or "dative of experience" remains the same when the contents of our conscious inner life become re-integrated into a broader context. The "eternal world-eye" is *literally* eternal.

But Schopenhauer also acknowledged that, while alive, we all have a second mode of existence corresponding to our physical body—that is, the extrinsic appearance or representation of our individual, dissociated contents of consciousness. This second mode, of course, will not survive death. Our narrative of self-identity will be seen through, just as we see through the identity of our dream avatars upon waking up.

This, in a nutshell, is analytic idealism, a philosophical view whose origins can be traced back—through Carl Jung (cf. Kastrup 2021), Arthur Schopenhauer (cf. Kastrup 2020), Emanuel Swedenborg, Parmenides (see Chapter 31 of this book) and many others—to the origins of civilization itself.

Chapter 14

Could Multiple Personality Disorder Explain Life, the Universe and Everything?

The condition now known as 'dissociative identity disorder' might help us understand the fundamental nature of reality

With Adam Crabtree and Edward F. Kelly

(The original version of this essay was published on Scientific American *on 18 June 2018)*

In 2015, doctors in Germany (Strasburger & Waldvogel) reported the extraordinary case of a woman who suffered from what has traditionally been called 'Multiple Personality Disorder' and today is known as 'Dissociative Identity Disorder' (DID). The woman exhibited a variety of dissociated personalities ('alters'), some of which claimed to be blind. Using EEGs, the doctors were able to ascertain that the brain activity normally associated with sight wasn't present while a blind alter was in control of the woman's body, even though her eyes were open. Remarkably, when a sighted alter assumed control, the usual brain activity returned.

This was a compelling demonstration of the *literally blinding* power of extreme forms of dissociation, a condition in which the psyche gives rise to multiple, operationally separate centers of consciousness, each with its own private inner life and memories.

Modern neuroimaging techniques have demonstrated that DID is real: in a 2014 study, doctors performed functional brain scans on both DID patients and actors simulating DID (Schlumpf *et al.*). The scans of the actual patients displayed clear differences when compared to those of the actors, showing that dissociation has an identifiable neural activity fingerprint. In other words, there is something rather particular that dissociative processes *look like* in the brain.

There is also compelling clinical data showing that different alters can be concurrently conscious and see themselves as distinct identities. One of us has written an extensive treatment of evidence for this distinctness of identity and the complex forms of interactive memory that accompany it, particularly in those extreme cases of DID that are usually referred to as multiple personality disorder (Crabtree 2009).

The history of this disorder dates back to the early 19th century, with a flourishing of cases in the 1880s through the 1920s, and again from the 1960s to the late-1990s. The massive literature on the subject confirms the consistent and uncompromising sense of separateness experienced by the alter personalities. It also displays compelling evidence that the human psyche is constantly active in producing personal units of perception and action that might be needed to deal with the challenges of life.

Although we may be at a loss to explain precisely how this creative process occurs—because it unfolds almost totally beyond the reach of self-reflective introspection—the clinical evidence nevertheless forces us to acknowledge something is happening that has important implications for our views about what is and is not possible in nature.

Now, one of us (Kastrup 2018a) posits that dissociation can offer a solution to a critical problem in our current understanding of the nature of reality. This requires some background, so bear with us.

According to the mainstream metaphysical view of materialism, reality is fundamentally constituted by physical stuff outside and independent of mind. Mental states, in turn, should be explainable in terms of the parameters of physical processes in the brain.

A key problem of materialism, however, is its inability to make sense of how our subjective experience of qualities—what it is like to feel the warmth of fire, the taste of an apple, the bitterness of disappointment, etc.—could arise from mere arrangements of physical stuff. Whereas physical entities such as subatomic particles possess abstract relational properties—such as mass, spin, momentum and charge—there is nothing about these properties, or the way particles are arranged in a brain, in terms of which one could deduce what the warmth of fire, the taste of an apple or the bitterness of disappointment feel like. This is known as the 'hard problem of consciousness' (Chalmers 2003).

To circumvent this problem, some philosophers have proposed an alternative: that experience is inherent to every fundamental physical entity in nature. Under this view—called 'constitutive panpsychism'—matter already has experience from the get-go, not just when it arranges itself in the form of brains. Even subatomic particles are supposedly subjects of experience, possessing some very simple form of consciousness. Our own human consciousness is then allegedly constituted by a combination of the subjective inner lives of the countless physical particles that make up our nervous system.

However, constitutive panpsychism has a critical problem of its own: there is arguably no coherent, non-magical way in which lower-level subjective points of view—such as those of subatomic particles or neurons in the brain—could combine to form higher-level subjective points of view, such as yours and ours. This is called the 'combination problem' and it is arguably just as insoluble as the hard problem of consciousness (Chalmers 2016a).

The obvious way around the combination problem is to posit that, although consciousness is indeed fundamental in nature, it isn't fragmented like matter. The idea is to extend consciousness to the entire fabric of spacetime, as opposed to limiting it to the boundaries of individual subatomic particles. Under this view—called 'cosmopsychism' in modern philosophy, although our preferred formulation of it boils down to what has classically been called 'idealism'—there is only one, universal, consciousness. The physical universe as a whole is the extrinsic appearance of universal inner life, just as a living brain and body are the extrinsic appearance of a person's inner life.

You don't need to be a philosopher to realize the obvious problem with this idea: people have private, separate fields of experience. We can't normally read your thoughts and, presumably, neither can you read ours. Moreover, we are not normally aware of what's going on across the universe and, presumably, neither are you. So for idealism to be tenable, one must explain—at least in principle—how one universal consciousness gives rise to multiple, private, but concurrently conscious centers of cognition, each with a distinct personality and sense of identity.

And here is where dissociation comes in. We know empirically from DID that consciousness can give rise to many operationally distinct centers of concurrent experience, each with its own personality and sense of identity. Therefore, if something analogous to DID happens at a universal level, the one universal consciousness could, as a result, give rise to many 'alters' with private inner lives like yours and ours. As such, we may all be 'alters'—dissociated personalities—of universal consciousness.

Moreover, as we've seen earlier, there is something dissociative processes look like in the brain of a patient with DID. So if some form of universal-level DID happens, the 'alters' of universal consciousness must also have an extrinsic appearance. We posit that this appearance is *life itself*: metabolizing organisms are simply what universal-level dissociative processes *look like*.

Idealism is a tantalizing view of the nature of reality, in that it elegantly circumvents two arguably insoluble problems: the hard problem of consciousness and the combination problem. Insofar as dissociation offers a path to explaining how, under idealism, one universal consciousness can become many individual minds, we may now have at our disposal an unprecedentedly coherent and empirically grounded way of making sense of life, the universe and everything.

Chapter 15

The Unexpected Origin of Matter

The external world is constituted by transpersonal experiential states

(The original version of this essay was published on IAI News *on 1 March 2020)*

The 'hard problem of consciousness' (Chalmers 2003) is not a problem that needs to be solved, for it doesn't exist in any objective sense. It is merely an internal contradiction of the reasoning behind metaphysical materialism, a conceptual short-circuit that arises as we logically work out the implications of the materialist conception of matter. There is no heroic challenge to be faced here, merely an embarrassing sign that our most basic assumptions about the nature of reality are dead wrong.

Like the rest of us, metaphysical materialists start from the contents of their own consciousness, such as perceptual experiences. All they are ever directly acquainted with are the colors, flavors and tones they perceive. But in order to explain why the external world we inhabit doesn't comply with our inner wishes and fantasies, materialists *consciously* postulate that the world is constituted by a medium outside and independent of consciousness—namely, matter. As such, matter is a tentative explanatory abstraction, a conceptual creation of reasoning consciousness. We can never become directly acquainted with matter, for all we ever know about the world are our conscious perceptions.

Having conjured up matter, materialists then posit that their consciousness—where matter is conceived to begin with—must be reducible *to* matter; that is, to one of consciousness's own abstractions. In a significant sense, this is a circular and self-defeating proposition, specifically because of how materialists define matter.

Indeed, under mainstream materialism matter is defined in purely *quantitative* terms: measurable values of mass, electric charge, momentum, position, frequency, amplitude, etc. Once these numerical values are determined—be it through direct measurement or inference—they ostensibly say *everything* there is to be said about matter; nothing else is left. There is nothing about matter that isn't captured by a list of numbers. Hence, under materialism matter—by definition—has only quantitative properties.

Where does this idea of using quantities to define the world come from? It's not difficult to see: quantities are very useful for describing the relative differences of the contents of perception. For instance, the relative difference between red and blue can be compactly described by frequency values: blue has a higher frequency than red, so we can quantify the visual difference between the two colors by subtracting one frequency from the other. But frequency numbers cannot absolutely describe a color: if you tell a congenitally blind person that red is an electromagnetic field vibration of about 430 THz, the person will still have no idea of what it feels like to see red. Quantities are useful for describing relative differences between qualities *already known experientially*, but they completely miss *the qualities themselves*.

And here is where materialism incurs its first fatal error: it replaces the qualitative world of colors, tones and flavors—the only external world we are directly acquainted with—with a purely quantitative description that structurally fails to capture any quality whatsoever. It mistakes the usefulness of quantities in determining relative differences between qualities for—absurdly—something that can replace the qualities themselves.

Next, materialism attempts to deduce the contents of

consciousness from the matter in our brain. In other words, it tries to recover the qualities of experience from mere quantities that, *by deliberate definition*, leave out everything that has anything to do with qualities in the first place. The self-defeating nature of this maneuver is glaringly obvious once one actually understands the magic materialism is trying to perform. This is precisely why the hard problem isn't just hard, but impossible *by construction*. Yet, instead of realizing this, we get lost in conceptual confusion and hope to, one day, heroically prevail against the hard problem. It would be an inspiring story of human resolve if it weren't so embarrassingly silly.

In summary, from within their own consciousness materialists fantasize about a world of matter putatively outside consciousness. This imagined world is, by deliberate definition, incommensurable with the qualities of conscious experience to begin with. Then, in a majestic feat of conceptual masochism, materialists set out to reduce the contents of consciousness to such an abstract ... well, content of consciousness. This is the tragicomic background story of the hard problem; a problem that need not be solved as much as *seen through* in all its gloriously self-defeating contradictoriness.

"But what is the alternative?" I hear you ask. If matter is a self-defeating concept, how can we explain the fact that we all seem to inhabit a common external world, whose dynamisms are clearly independent of our own conscious inner life?

First of all, let me immediately acknowledge the empirically obvious: there is a world beyond and independent of our *individual* consciousness; a world that we all inhabit. And, alas, we clearly can't change how this world works by a mere act of individual conscious volition. But to acknowledge this does not require the bankrupt notion of matter outside consciousness. It only requires a *transpersonal* consciousness within which our *individual* consciousnesses are immersed.

Indeed, I maintain that the external world is itself constituted by transpersonal experiential states that simply present themselves to us in the form we call 'matter.' As such, 'matter' is merely the extrinsic appearance—the image—of inner experience; there's nothing more to it. In the case of living beings, the 'matter' constituting their body is the extrinsic appearance of their *individual* experiential states (this being the reason why measurable patterns of brain activity correlate with inner experience). In the case of the inanimate universe, on the other hand, 'matter' is the extrinsic appearance of *transpersonal* experiential states.

This hypothesis completely circumvents the hard problem, for it doesn't require reducing qualities to mere quantities; it only requires reducing certain qualities *to other qualities*: the colors, tones and flavors that appear on our screen of perception are modulated by transpersonal, endogenous experiential states—such as instinctive thoughts and feelings—underlying the inanimate universe around us. This modulation between different types of qualities happens every day in our own consciousness: our thoughts routinely modulate our emotions, and vice versa, although thoughts and emotions are qualitatively very different.

I am keenly aware that what I am suggesting above raises many questions, but it is impossible to answer them in a short essay such as this. Indeed, I've spent over a decade trying to give these questions a proper treatment, which required the many books and articles in my body of work. Here, beyond making clear that the hard problem is merely an internal contradiction of a bankrupt system of thought, I just wanted to hint at how it can be circumvented if one … well, thinks straight.

Part V

On Physics

Chapter 16

Should Quantum Anomalies Make Us Rethink Reality?

Inexplicable lab results may be telling us we're on the cusp of a new scientific paradigm

(The original version of this essay was published on Scientific American *on 19 April 2018)*

Every generation tends to believe that its views on the nature of reality are either true or quite close to the truth. We are no exception to this: although we know that the ideas of earlier generations were each time supplanted by those of a later one, we still believe that this time we got it right. Our ancestors were naïve and superstitious, but we are objective—or so we tell ourselves. We know that matter/energy, outside and independent of mind, is the fundamental stuff of nature, everything else being derived from it—Or do we?

In fact, studies have shown that there is an intimate relationship between the world we perceive and the conceptual categories encoded in the language we speak. We don't perceive a purely objective world out there, but one subliminally pre-partitioned and pre-interpreted according to culture-bound categories. For instance, "color words in a given language shape human perception of color" (Adelson 2005). A brain imaging study suggests that language processing areas are directly involved even in the simplest discriminations of basic colors (Tan *et al.* 2008). Moreover, this kind of "categorical perception is a phenomenon that has been reported not only for color, but for other perceptual continua, such as phonemes, musical tones and facial expressions" (Roberson & Hanley 2007). In an important sense, we see what our unexamined cultural categories teach us to see, which may help explain why every generation is so confident in their own worldview. Allow me to elaborate.

The conceptual-ladenness of perception isn't a new insight. Back in 1957, philosopher Owen Barfield wrote:

> I do not perceive any *thing* with my sense-organs alone. … Thus, I may say, loosely, that I 'hear a thrush singing.' But in strict truth all that I ever merely 'hear'—all that I ever hear simply by virtue of having ears—is sound. When I 'hear a thrush singing,' I am hearing … with all sorts of other things like mental habits, memory, imagination, feeling and … will. (Barfield 2011: 20, original emphasis)

As argued by philosopher Thomas Kuhn in his seminal book *The Structure of Scientific Revolutions* (2012), science itself falls prey to this inherent subjectivity of perception. Defining a "paradigm" as an "implicit body of intertwined theoretical and methodological belief," he wrote:

> something like a paradigm is prerequisite to perception itself. What a man sees depends both upon what he looks at and also upon what his previous visual-conceptual experience has taught him to see. In the absence of such training there can only be, in William James's phrase, "a bloomin' buzzin' confusion." (Kuhn 2012: 113)

Hence, because we perceive and experiment on things and events partly defined by an implicit paradigm, these things and events tend to confirm, *by construction*, the paradigm. No wonder, then, that we are so confident today that nature consists of arrangements of matter/energy outside and independent of mind: matter/energy is what we see when we look at the world.

Yet, as Kuhn pointed out, when enough "anomalies"—

empirically undeniable observations that cannot be accommodated by the reigning belief system—accumulate over time and reach critical mass, paradigms change. We may be close to one such a defining moment today, as an increasing body of evidence from quantum mechanics (QM) renders the current paradigm untenable.

Indeed, according to the current paradigm, the observable properties of an object should exist and have definite values even when the object is not being observed: the moon should exist and have whatever weight, shape, size and color it has even when nobody is looking at it. Moreover, a mere act of observation should not change the values of these properties. Operationally, all this is captured in the notion of 'non-contextuality': the outcome of an observation should not depend on the way other, separate but simultaneous observations are performed. After all, what I perceive when I look at the night sky should not depend on the way other people look at the night sky along with me, for the properties of the night sky uncovered by my observation should not depend on theirs.

The problem is that, according to QM, the outcome of an observation *can*—and indeed, often *does*—depend on the way another, separate but simultaneous, observation is performed. This happens with so-called 'quantum entanglement' and it contradicts the current paradigm in an important sense, as discussed above. Although Einstein argued that the contradiction arose merely because QM is incomplete (Einstein, Podolsky & Rosen 1935), John Bell proved mathematically that the predictions of QM regarding entanglement cannot be accounted for by Einstein's alleged incompleteness (1964).

So to salvage the current paradigm there is an important sense in which one has to reject the predictions of QM regarding entanglement. Yet, since Alain Aspect's seminal experiments in 1981-82 (Aspect, Grangier & Roger 1981, 1982, Aspect, Dalibard & Roger 1982), these predictions have been repeatedly confirmed, with potential experimental loopholes closed one by one. 1998 was a particularly fruitful year, with two remarkable experiments performed in Switzerland (Tittel *et al.*) and Austria (Weihs *et al.*). In 2011 (Lapkiewicz *et al.*) and 2015 (Manning *et al.*), new experiments again challenged non-contextuality. Commenting on this, physicist Anton Zeilinger has been quoted as saying that "there is no sense in assuming that what we do not measure [that is, observe] about a system has [an independent] reality" (Ananthaswamy 2011). Finally, Dutch researchers successfully performed a test closing all remaining potential loopholes (Hensen *et al.* 2015), which was considered by *Nature* magazine the "toughest test yet" (Merali 2015).

The only alternative left for those holding on to the current paradigm is to postulate some form of non-locality: nature must have—or so they speculate—observation-independent hidden properties, entirely missed by QM, which are 'smeared out' across spacetime. It is this allegedly omnipresent, invisible but objective background that supposedly orchestrates entanglement from 'behind the scenes.'

It turns out, however, that some predictions of QM are incompatible with non-contextuality even for a large and important class of non-local theories (Leggett 2003). Experimental results reported in 2007 (Gröblacher *et al.*) and 2010 (Romero *et al.*) have confirmed these predictions. To reconcile these results with the current paradigm would require a profoundly counterintuitive redefinition of what we call 'objectivity.' And since contemporary culture has come to associate objectivity with reality itself, the science press felt compelled to report on this by pronouncing, "Quantum physics says goodbye to reality" (Cartwright 2007).

The tension between the anomalies and the current paradigm can only be tolerated by *ignoring* the anomalies. This has been possible so far because the anomalies are only observed in laboratories. Yet *we know that they are there*, for their existence has been confirmed beyond reasonable doubt. Therefore, when we believe that we see objects and events outside and independent of mind, we are wrong in at least one very important sense. A new paradigm is needed to accommodate and make sense of the anomalies; one wherein mind itself is understood to be the essence—cognitively but also physically—of what we perceive when we look at the world around ourselves.

Chapter 17

Coming to Grips with the Implications of Quantum Mechanics

The question is no longer whether quantum theory is correct, but what it means

With Henry P. Stapp and Menas C. Kafatos

(The original version of this essay was published on Scientific American *on 29 May 2018)*

For almost a century, physicists have wondered whether the most counterintuitive predictions of quantum mechanics (QM) could actually be true. Only in recent years has the technology necessary for answering this question become accessible, enabling a string of experimental results—including startling ones reported in 2007 (Gröblacher *et al.*) and 2010 (Romero *et al.*), and culminating now with a remarkable test (BIG Bell Test Collaboration 2018)—showing that key predictions of QM are indeed correct. Taken together, these experiments indicate that the everyday world we perceive does not exist until observed, which in turn suggests—as we shall argue in this essay—a primary role for mind in nature. It is thus high time the scientific community at large—not only those involved in foundations of QM—faced up to the counterintuitive implications of QM's most controversial predictions.

Over the years, we have written extensively about why QM seems to imply that the world is essentially mental (Kafatos & Nadeau 1990, Nadeau & Kafatos 1999, Stapp 1993, 2001, 2007, 2017, Kastrup 2019). We are often misinterpreted—and misrepresented—as espousing solipsism or some form of 'quantum mysticism,' so let us be clear: our argument for a mental world does not entail or imply that the world is merely one's own personal hallucination or act of imagination. Our view is entirely naturalistic: the mind that underlies the world is a transpersonal mind behaving according to natural laws. It comprises but far transcends any individual psyche.

The claim is thus that the dynamics of all inanimate matter in the universe correspond to *trans*personal mentation, just as an individual's brain activity—which is also made of matter—corresponds to *personal* mentation. This notion eliminates arbitrary discontinuities and provides the missing inner essence of the physical world: all matter—not only that in living brains—is the outer appearance of inner experience, different configurations of matter reflecting different patterns or modes of mental activity.

According to QM, the world exists only as a cloud of simultaneous, overlapping possibilities—technically called a 'superposition'—until an observation brings one of these possibilities into focus in the form of definite objects and events. This transition is technically called a 'measurement.' One of the keys to our argument for a mental world is the contention that only conscious observers can perform measurements.

Some criticize this contention by claiming that inanimate objects, such as detectors, can also perform measurements, in the sense described above. The problem is that the partitioning of the world into discrete inanimate objects is merely nominal. Is a rock integral to the mountain it helps constitute? If so, does it become a separate object merely by virtue of its getting detached from the mountain? And if so, does it then perform a measurement each time it comes back in contact with the mountain, as it bounces down the slope? Brief contemplation of these questions shows that the boundaries of a detector are arbitrary. The inanimate world is a single physical system governed by QM. Indeed, as first argued by John von Neumann (1996) and rearticulated in the work of one of us (Stapp 2001), when two inanimate objects interact they simply become quantum mechanically 'entangled' with one another—that is, they become united in such a way that the behavior of one becomes inextricably linked to the behavior of the other—but no actual measurement is performed.

Let us be more specific. In the well-known double-slit experiment, electrons are shot through two tiny slits. When they are observed at the slits, the electrons behave as definite particles. When observed only after they've passed through the slits, the electrons behave as clouds of possibilities. In 1998, researchers at the Weizmann Institute in Israel showed that, when detectors are placed at the slits, the electrons behave as definite particles (Buks *et al.*). At first sight, this may seem to indicate that measurement does not require a conscious observer.

However, *the output of the detectors only becomes known when it is consciously observed by a person*. The hypothesis of a measurement before this conscious observation lacks compelling theoretical or empirical grounding. After all, as discussed above, QM offers no reason why the whole system—electrons, slits and detectors combined—wouldn't be in an entangled superposition before someone looks at the detectors' output.

As such, a conscious observer may be indispensable, an idea further elaborated by one of us with regard to so-called 'delayed choice quantum eraser' experiments (Narasimhan & Kafatos 2016). The bottom line is that we cannot know that detectors actually perform measurements, for we cannot abstract ourselves out of our knowledge. Recall Max Planck's position: "I regard consciousness as fundamental. I regard matter as derivative from consciousness. *We cannot get behind consciousness*" (emphasis added).

Some claim that the modern notion of 'decoherence' rules out consciousness as the agency of measurement. According to this claim, when a quantum system in a superposition state is probed, information about the overlapping possibilities in the superposition 'leaks out' and becomes dispersed in the surrounding environment (Zurek 2003). This allegedly explains in a fairly mechanical manner why the superposition becomes indiscernible after measurement.

The problem, however, is that decoherence cannot explain how the state of the surrounding environment becomes definite to begin with, so it doesn't solve the measurement problem or rule out the role of consciousness. Indeed, as Wojciech Zurek—one of the fathers of decoherence—admitted,

> ... an exhaustive answer to [the question of why we perceive a definite world] would undoubtedly have to involve a model of 'consciousness,' since what we are really asking concerns our [observers'] impression that 'we are conscious' of just one of the alternatives. (1994)

As a matter of fact, peculiar statistical characteristics of the behavior of entangled quantum systems—namely, their experimentally confirmed violation of so-called Bell's (Bell 1964) and Leggett's (Leggett 2003) inequalities—seem to rule out everything but consciousness as the agency of measurement.

Some then claim that entanglement is observed only in microscopic systems and, therefore, its peculiarities are allegedly irrelevant to the world of tables and chairs. But such a claim is untrue, as several recent studies (Ansmann *et al.* 2009, Lee *et al.* 2011, Klimov *et al.* 2015) have demonstrated entanglement for much larger systems. Recently, a paper reported entanglement even for "massive" objects (Ockeloen-Korppi *et al.* 2018). Moreover, quantum superposition has been observed in systems as varied as small metal paddles (O'Connell *et al.* 2010) and living tissue (Engel *et al.* 2007). Clearly, the laws of QM apply at all scales and substrates.

What preserves a superposition is merely how well the quantum system—whatever its size—is isolated from the world of tables and chairs known to us through direct *conscious apprehension*. That a superposition does not survive exposure to this world suggests, if anything, a role for consciousness in the emergence of a definite physical reality.

Now that the most philosophically controversial predictions of QM have—finally—been experimentally confirmed without remaining loopholes, there are no excuses left for those who want to avoid confronting the implications of QM. Lest we continue to live according to a view of reality now known to be false, we must shift the cultural dialogue towards coming to grips with what nature is repeatedly telling us about herself.

Chapter 18

Reasonable Inferences From Quantum Mechanics

A Response to "Quantum Misuse in Psychic Literature"

(The original version of this academic paper was published in the Journal of Near-Death Studies, *volume 37, issue 3, fall of 2019, pages 185-200, DOI: 10.17514/JNDS-2019-37-3-p185-200.)*

Abstract: This invited article is a response to the paper "Quantum Misuse in Psychic Literature," by Jack A. Mroczkowski and Alexis P. Malozemoff, published in the fall-2019 issue of the *Journal of Near-Death Studies*. Whereas I sympathize with Mroczkowski's and Malozemoff's cause and goals, and I recognize the problem they attempted to tackle, I argue that their criticisms often overshot the mark and end up adding to the confusion. I address nine specific technical points that Mroczkowski and Malozemoff accused popular writers in the fields of health care and parapsychology of misunderstanding and misrepresenting. I argue that, by and large—and contrary to Mroczkowski's and Malozemoff's claims—the statements made by these writers are often reasonable and generally consistent with the current state of play in foundations of quantum mechanics.

I appreciate this opportunity to respond to Jack A. Mroczkowski and Alexis P. Malozemoff's (2019) article "Quantum Misuse in Psychic Literature" published in the fall-2019 issue of the *Journal of Near-Death Studies*. Let me start by acknowledging that I sympathize with Mroczkowski's and Malozemoff's—henceforth 'the authors'—cause. Few scholars would deny that Quantum Mechanics (QM) has been the subject of misuse, so it is laudable that these authors have attempted to correct at least some of it. The authors' goal of encouraging others "to avoid augmenting their discussions with improper references to physics" (p. 132) is unimpeachable and timely. There is much I agree with in their paper.

However, I believe the authors overshot the mark with their criticism. In my view, some of what they considered misuse are legitimate —if sometimes poorly worded—attempts to highlight that QM defies most people's ordinary prejudices about the nature of reality. These prejudices define what is typically considered plausible or implausible, thereby motivating many people to mistakenly dismiss important possibilities in fields such as health care and parapsychology.

Although QM has been around for nearly a century, its implications haven't yet percolated through other scientific disciplines. As a matter of fact, even within physics itself, the community of 'foundations of physics'—scholars who ponder the metaphysical implications of QM—is relatively small. From this perspective, it is difficult—at least in principle—to fault attempts to bring to popular attention the degrees of natural freedom that QM may open up.

Bizarrely, popular culture is still dominated by the constraints of a naïve local-realism that QM has definitively relegated to the trash bin of history. It is thus not only legitimate, but arguably even imperative, that thought leaders play a prominent role in expanding cultural horizons in this regard. The formidable momentum behind naïve local-realism must be countered, lest people continue to live under a limiting and—most importantly—mistaken view of reality.

Although some popular writers may have worded their claims inaccurately, discussing the mind-bending implications of QM both accessibly and accurately is a formidable challenge. The authors themselves—who, unlike most of the popular writers they criticized, have the advantage of being experts in the field—have been admittedly unable to do so. They compensated for this shortcoming by adding parenthetical clarifications accessible only to experts. Yet, as far as regular readers are concerned, these parenthetical clarifications do nothing to prevent misunderstandings; they merely serve as disclaimers to safeguard the authors.

The impasse readers are left with is thus the following: on the one hand, thought leaders in fields where the implications of QM are salient must engage their public on the possibilities these implications open up; on the other hand, they must do it in a minimally accurate manner. There is significant tension between these two goals and no magic bullet to resolve it. The best way forward may be to engage in a critical dialogue in which scholars seek to find a balance. This is the spirit of the present response.

In the next sections, I shall comment on nine specific technical points raised by the authors.

18.1 The physical world as illusion

The authors suggested that the characterization of the physical world as illusory is not justified by QM. To evaluate whether this suggestion is correct requires first an understanding of what it means to claim that the world is an illusion.

Most ordinary people would take the world to be real—as opposed to illusory—if its measurable physical properties existed independently of whether and how they are observed. An act of observation should merely *disclose* a self-existing physical reality, not create or define it. This presumed independence from observation—technically called 'non-contextuality'—is what underlies most people's intuition of the world's concreteness. To claim that the world is an illusion therefore means to deny non-contextuality: if the physical properties of the world actually depend on how they are observed—as opposed to existing in and by themselves—then the world is an illusion.

So what does QM say about it? Operationally, non-contextuality means that the outcome of a measurement should not depend on the way another, separate but simultaneous, measurement is performed. According to quantum theory, however, this is simply not the case. The relevant question is then whether quantum theory is correct.

Since Alain Aspect's seminal experiments (Aspect, Dalibard & Roger 1982, Aspect, Grangier & Roger 1981, 1982), the predictions of quantum theory in this regard have been repeatedly confirmed. The year 1998 was particularly fruitful, with two remarkable experiments performed in Switzerland (Tittel *et al.*) and Austria (Weihs *et al.*). More recent experiments again challenged non-contextuality (Lapkiewicz *et al.* 2011, Manning *et al.* 2015). Commenting on them, physicist Anton Zeilinger has been quoted as saying that "there is no sense in assuming that what we do not measure about a system has [an independent] reality" (Ananthaswamy 2011). Finally, Dutch researchers (Hensen *et al.* 2015) and a large international collaboration (BIG Bell Test Collaboration 2018) successfully performed tests closing all potential loopholes and definitively proving quantum theory correct.

The only way out for the adherents of non-contextuality is to speculate about the existence of hidden physical properties 'smeared out' across spacetime. It turns out, however, that certain predictions of quantum theory are incompatible with non-contextuality even for a large and important class of such speculations (Leggett 2003). Experiments have now confirmed these predictions (Gröblacher *et al.* 2007, Romero *et al.* 2010) with results so significant that the science press has felt compelled to pronounce, "Quantum physics says goodbye to reality" (Cartwright 2007).

The surviving interpretation of QM that could, in principle, still preserve non-contextuality is Bohmian Mechanics (Bohm 1952a, 1952b). Alas, this interpretation is plagued by a number of other problems. For instance, unlike regular QM with its

Quantum Field Theory extensions, Bohmian Mechanics has no relativistic version. Physicists Raymond Streater and Luboš Motl have reviewed other compelling technical arguments against Bohmian Mechanics (Motl 2009, Streater 2007: 103-112). Finally, recent experiments have reportedly refuted the interpretation empirically (Wolchover 2018).

Admittedly, there is still polemic surrounding not only Bohmian Mechanics but also the experimental results that refute non-contextuality. It is nonetheless fair to say that never before has the idea of a real physical world, independent of observation, looked so precarious. Non-contextuality, if not dead, is in serious trouble.

Consequently, it seems entirely reasonable to claim that, as far as QM is concerned, the physical world people ordinarily experience is indeed akin to an 'illusion.' Prior to being observed, observable physical quantities are only potentials—modeled by waves of probabilities—as opposed to defined existences.

18.2 Personal physical realities

The authors criticized the assertion by Deepak Chopra that "the physical world, including our bodies, is a response of the observer. We create our bodies as we create the experience of our world" (as quoted in Mroczkowski & Malozemoff 2019: 144). Chopra went on to acknowledge that "these are vast assumptions, the makings of a new reality, yet all are grounded in the discoveries of quantum physics made almost a hundred years ago" (*Ibid.*). So, once again, the question remains whether grounding—not proof, not irrefutable evidence, just grounding for this claim of Chopra's—exists within QM.

As the authors acknowledged, QM has many different metaphysical interpretations. There is no consensus in physics regarding which interpretation is more likely, let alone true. But one of the more sober, parsimonious and flat-out honest interpretations is Carlo Rovelli's (1996) Relational Quantum

Mechanics (RQM). According to RQM, there are no absolute—that is, observer-independent—physical quantities. Instead, *all* physical quantities—the entire physical world—are relative to the observer in a way analogous to motion.

Rovelli summarized RQM thus:

> [Because] different observers give different accounts of the same sequence of events, ... each quantum mechanical description has to be understood as relative to a particular observer. Thus, a quantum mechanical description of a certain system (state and/or values of physical quantities) cannot be taken as an 'absolute' (observer-independent) description of reality, but rather as a formalization, or codification, of properties of a system relative to a given observer. (1996: 1648)

The implication is that each person, as an individual observer, 'inhabits' his or her own physical world, as defined by the context of his or her own observations. This assertion comes very close to the notion, suggested by Chopra, that each person lives in a physical reality created in response to one's own observations.

However, a reader might inquire whether RQM is true. Any definite answer to this question would overlook the morass of unending discord that prevails in the field of foundations of physics. Nevertheless, a very recent and significant experimental result has arguably proven the central and defining point of RQM: that the physical world is, indeed, *relative to the observer* in a way analogous to motion (Proietti *et al.* 2019, Emerging Technology from the arXiv 2019).

Therefore, in view of the current state of play in QM, Chopra's statements—albeit speculative—are neither crazy nor ungrounded in QM. Counterintuitive as it may sound, the idea of relative physical worlds can even be reconciled with the experience that all people share a common environment. I address this idea more fully in Chapter 6 of my book, *The Idea of the World* (Kastrup 2019: 93-122).

18.3 Choice and randomness

The authors criticized popular writers who suggested that intention may directly influence the physical world's transition from potentials to defined existences—that is, the so-called 'collapse of the wave function' somehow associated with an act of observation. The authors reasoned that, according to QM, collapse produces random outcomes, thereby "preventing a person from choosing or intending a particular desirable outcome" (Mroczkowski & Malozemoff 2019: 137).

I believe the authors' reasoning here is flawed. First, it is important to consider that randomness is a highly ambiguous concept: whereas it is defined as the absence of recognizable patterns or biases—there being formal randomness tests to verify whether this is the case—a truly random process can, theoretically, produce *any* pattern. The chance of finding a pattern in a truly random process may be small, but it isn't zero. Indeed, because it basically consists of an acknowledgment of causal ignorance, randomness is an extraordinarily accommodating notion.

Given this point, insisting that a process is random doesn't actually exclude any outcome whatsoever. It is physically coherent—whether plausible or not—that intention may indeed influence collapse outcomes without violating quantum theory. To argue that this cannot be the case merely because QM does not *positively* predict such an effect begs the question: the point in contention is precisely that there may be natural agencies or organizing principles that current science still fails to recognize.

Moreover, the randomness of wave function collapse is defined on the basis of a series of repeated observations of the same quantum system. For instance, if one measures the spin of an electron along a certain direction, the result will be either

½ or –½. If one then resets the experiment—ensuring that all initial conditions are the same as before—and redoes the measurement, again the result will be either ½ or –½. A series of such measurements will produce a string of numbers. *It is the string that should meet randomness criteria.*

But a technically random string, of course, does not preclude the possibility that *individual measurements* within it can be influenced by intention in a way that may not be noticeable in the overall string. And even if an overall statistical bias is noticeable, no skeptic will raise an eyebrow as, theoretically, random processes can—as argued above—produce *any* conceivable pattern by mere chance. This renders the authors' assertion effectively non-falsifiable.

18.4 Synchronicity

Quantum predictions hold only at a statistical level. The outcomes of individual measurements—that is, individual observations or events—are *non*-deterministic and *un*predictable; quantum theory enforces no result whatsoever at the level of individual outcomes. It is this causally undetermined space that psychiatrist Carl Jung and Nobel Prize Laureate physicist Wolfgang Pauli populated with their notion of 'synchronicity': acausal meaningful coincidences that allegedly reflect archetypal patterns underlying not only the human psyche, but also the physical world at large (Jung 1985, Jung & Pauli 2001).

The authors denied "that something about quantum theory may explain the serial coincidences that underlie synchronicity" (Mroczkowski & Malozemoff 2019: 150). I believe this statement, although strictly correct, is misleading in that it sets up a straw man. The point is not that QM *positively* accounts for synchronicities; the point is that—*unlike classical physics*—QM *leaves space open* for synchronicities. Indeed, according to QM, at its most fundamental level nature is not deterministic; there is no causal necessity enforced at that level. This notion opens the door to other organizing principles still unknown to science.

The authors repeatedly argued that wave function collapse leads to random outcomes. But this seeming randomness does not contradict synchronicity either: regarding the latter, theorists have postulated that nature organizes itself according to global archetypal patterns. These *global* patterns can be easily reconciled with apparent randomness at the level of *individual* quantum events, as I illustrate with the following analogy.

Imagine that you toss three dice on a table, multiple times. After each toss, each individual die randomly displays a number from one to six. In other words, the behavior of each die is seemingly random from toss to toss. But now imagine that, when you look at all three dice together, after every toss, you realize that either they *all* display an even number or they *all* display an odd number. This is a simple hypothetical example of a global, synchronistic pattern that can occur even when the individual constituent events, considered in isolation, meet randomness criteria. In Jung's words, "Within the randomness of the throwing of the dice, a 'psychic' orderedness comes into being" (Jung & Pauli 2001: 62).

If this kind of global synchronistic alignment were to happen across quantum events in the world at large, physicists would be none the wiser. For although they can test individual events in the laboratory and verify that, when taken in isolation, the events are random, they wouldn't be able to discern a *global* pattern within the complexity of the physical world at large; there are just too many 'dice' to look at under controlled laboratory conditions.

The relationship between synchronicity and QM, which I articulated above, has been vouchsafed by Pauli himself. After reviewing the final draft of Jung's synchronicity essay, Pauli wrote: "I ... found that ... from the standpoint of modern physics, [the essay] is now unassailable" (Jung & Pauli 2001: 71).

18.5 Emptiness

Many popular writers have highlighted the fact that, when looked at closely, matter reveals itself to be mostly empty space. If one considers the total volume of an atom and compares it to the aggregate volume of its constituent mass-containing elementary subatomic particles—such as quarks and leptons—one realizes that the atom is indeed mostly empty. The authors, however, argued that such a conclusion "bears no relation to modern quantum physics" because "the wave function [of the subatomic particles] fills space" (Mroczkowski & Malozemoff 2019: 144).

I believe the authors' argument here is flawed in more ways than one. First, they seem to have implicitly taken for granted that the wave function is *ontic*—that it corresponds to an objectively existing physical entity smeared out across space. There is certainly no consensus in physics that this is the case. Many physicists maintain, instead, that the wave function is merely *epistemic*—that it merely captures the extent of human knowledge regarding nature's future behavior. If the latter position is true, then there is nothing objectively real that "fills space" inside an atom.

Be that as it may, when one says that an atom is 'mostly empty' one is referring to the fact that most of the space in the atom contains *no mass*. A well-illuminated vacuum is still considered empty, in that photons have no mass. Similarly, a vacuum filled with electromagnetic fields is still empty, for fields—abstract mathematical tools—do not count as 'occupants' of space as far as the popular intuition about 'emptiness' is concerned. Now, because mass is a measurable physical quantity—an 'observable'—one can speak of its existence only *after* wave function collapse—or whatever passes for collapse, because even that phenomenon is not consensus in physics today. What then remains is a set of mass-containing elementary subatomic particles that, indeed, occupy but a tiny fraction of the atom's total volume.

There is just no denying that 20th-century subatomic physics has ushered in an understanding that contradicts popular intuitions about the solidity of matter. These intuitions are a throwback to outdated Greek atomist views. In this context, I believe it to be valid that popular writers point out to their audiences that, contrary to vulgar assumptions, matter indeed is 'mostly empty space.'

18.6 Consciousness as the agency of collapse

The authors criticized the notion that consciousness might be the agency behind the transition of the physical world, upon observation, from mere potentialities to defined physical quantities. They said, "this interpretation has not been proven and is not generally accepted by quantum physicists" (Mroczkowski & Malozemoff 2019: 138). Although strictly correct, such statement also sets up a straw man: *no* interpretation of QM has been proven or generally accepted by quantum physicists. I do not believe this state of affairs should stop all authors from ever alluding to, or speculating about, the implications of QM.

If consciousness does not cause wave function collapse—or whatever passes for collapse—then it follows that an inanimate entity of some sort must be responsible for it. Yet, the claim that inanimate objects—such as electronic detectors—can perform quantum mechanical measurements is fundamentally problematic, because the partitioning of the world into discrete inanimate objects is merely nominal to begin with. Is a rock integral to the mountain it helps constitute? If so, does it become a separate object merely by virtue of its getting detached from the mountain? And if so, does it then perform a quantum measurement—that is, an observation that causes collapse of the wave function—each time it comes back in contact with the mountain as it bounces down the slope? Brief contemplation of these questions shows that the boundaries of a detector are arbitrary.

Indeed, as John von Neumann (1996) first argued, when two inanimate objects interact they simply become quantum mechanically entangled with one another—that is, they become united in such a way that the behavior of one becomes inextricably linked to the behavior of the other—but no actual measurement is performed. As such, the inanimate world is a unitary, indivisible physical system governed by QM. There are no detectors performing measurements; there is only the *one* inanimate world. In the words of Erich Joos, "because of the non-local properties of quantum states, a consistent description of some phenomenon in quantum terms must finally include the entire universe" (2006: 71).

Let me use a concrete example to be more specific. In the well-known double-slit experiment, electrons are shot through two tiny slits. When they are observed at the slits, the electrons behave as defined individual particles. But when observed only after they have passed through the slits, the electrons behave as superposed potentialities. In 1998, researchers at the Weizmann Institute in Israel showed that, when detectors are placed at the slits, the electrons behave as defined individual particles (Buks *et al.* 1998). At first sight, this result may seem to indicate that measurement does not require a conscious observer.

However, the output of the detectors becomes known only when it is *consciously observed* by a person. The hypothesis of a measurement before this conscious observation lacks compelling theoretical and empirical grounding. After all, QM offers no reason why the whole system—electrons, slits and detectors combined—should not be in an entangled superposition before and until someone looks at the detectors' output (von Neumann 1996). Its condition simply cannot be known. Because people cannot abstract themselves out of their knowledge, they cannot know that detectors actually perform measurements and cause wave function collapse.

Consequently, as far as people *can* know, before it is represented through conscious perception the world consists of a unitary superposition of potentialities. This superposition—indivisible, as quantum entanglement prevents elements of the superposition from being describable separately from one another—is incompatible with the existence of individual, separate objects and events with defined properties.

18.7 Decoherence

The authors maintained that a quantum phenomenon called 'decoherence' "is responsible for [destroying] most unique quantum interference effects, and this decoherence always happens before any conscious observation" (Mroczkowski & Malozemoff 2019: 138). They seemed to suggest that decoherence obviates the postulate that consciousness is the agency of collapse. In other words, the suggestion seems to have been that decoherence alone already explains the transition of a quantum system from mere potentialities to defined physical quantities. This, however, is a well-known fallacy.

If one takes for granted the existence of a macroscopic environment consisting of defined physical quantities—that is, a classical environment—from which a microscopic quantum system in a superposition is initially isolated, then it is true that any contact with the environment will destroy the superposition. Information about the overlapping potentialities will 'leak out' and become dispersed in the surrounding environment.

The problem, however, is that decoherence cannot explain how the state of the surrounding environment became defined—that is, classical—to begin with, so it doesn't solve the measurement problem or rule out the role of consciousness. As Wojciech Zurek—one of the fathers of decoherence—acknowledged,

> An exhaustive answer to [the question of why we perceive a classical world instead of superposed potentialities] would

> undoubtedly have to involve a model of 'consciousness,' since what we are really asking concerns our [observers'] impression that 'we are conscious' of just one of the alternatives. (1994: 29)

And as Joos pointed out, "the effects of decoherence just look like collapse" (2006: 77). Indeed, in an essay dedicated to highlighting the role of decoherence in the emergence of a classical world from a quantum substrate, Joos ultimately concluded that some form of either wave function collapse or parallel universes is still needed (*Ibid.*: 75). Decoherence alone will not do.

In conclusion, decoherence does not obviate or preclude the possibility that consciousness is the agency behind collapse.

18.8 Microscopic versus macroscopic

Another argument line frequently repeated by the authors is that the quantum phenomena popular writers rely on occur mostly at a microscopic level. The authors seemed to imply that fundamental or metaphysical conclusions cannot be extrapolated from this microscopic realm to the macroscopic world of tables and chairs.

Although there are undeniable operational differences between the behavior of the world of tables and chairs and that of isolated microscopic quantum systems, these differences cannot be fundamental; they must, instead, be merely epiphenomenal. After all, the world is quantum, for macroscopic objects and events are just compound results of microscopic dynamics. To quote Joos once again,

> [A] method for sweeping the interpretive problems under the carpet is simply to assume, or rather postulate, that quantum theory is only a theory of micro-objects, whereas in the macroscopic realm per decree (or should I say wishful thinking?) a classical description has to be valid. Such

> an approach leads to the endlessly discussed paradoxes of quantum theory. These paradoxes only arise because this particular approach is conceptually inconsistent ... In addition, micro- and macro-objects are so strongly dynamically coupled that we do not even know where the boundary between the two supposed realms could possibly be found. For these reasons it seems obvious that there is no boundary. (2006: 74-75)

He went on to say, "whichever interpretation [of QM] one prefers, the classical world view has been ruled out" (*Ibid.*: 76). It is this understanding that motivates popular writers to speculate about what new degrees of natural freedom may open up, in the macroscopic world, when the implications of QM are considered. There is nothing wrong—at least in principle—with this extrapolation, for there is no actual boundary between the microscopic and the macroscopic. The distinction between the two is arbitrary, nominal, motivated by convenience and purely epistemic.

18.9 Superluminal information transfer

The authors frequently alluded to the 'no-communication theorem' of quantum information theory to emphasize that quantum entanglement—despite its "spooky action at a distance"—cannot be used for faster-than-light information transfer. This conclusion is, of course, entirely correct. The problem is that the authors seemed to set up yet another straw man by implying that popular writers have relied on superluminal communication to account for psi phenomena. In this regard, they singled out Pim van Lommel's allusion to the notion of non-local consciousness.

The straw man here is as follows: the very idea of non-local consciousness entails that reality is fundamentally one and, as such, *communication is obviated* to begin with. In the words of

Jonathan Schaffer, "physically, there is good evidence that the cosmos forms an entangled system and good reason to treat entangled systems as irreducible wholes" (2010: 32).

Therefore, at the most fundamental level of reality—the level in which psi phenomena allegedly occur—there is no need for information transfer to begin with. In the altered state of consciousness near-death experiencers find themselves in, they are ostensibly one with all existence, and so the information in question is already 'in them,' so to speak. Nothing needs to be communicated from one place to another because the information is, *ex hypothesi*, already 'everywhere.' The allusion here is rather to something akin to Bohm's (1980) 'implicate order' than to information transfer, superluminal or otherwise.

Notice that I am not necessarily arguing for, or defending, psi, for I am not familiar enough with the subject to take an informed position either way. I am just pointing out that the argument for psi some popular writers make, though certainly reliant on the implications of QM, does not—contrary to what the authors claimed—necessarily entail superluminal communication. The allusion to entanglement is meant to underpin the possibility that the entire cosmos is fundamentally a *unitary whole*, not necessarily an appeal to information transfer through entanglement.

Commentary

A recurring theme in the authors' argumentation is the claim that certain views—particularly those related to foundations of physics—are not generally accepted by physicists, are controversial, disputed and so forth. The repeated suggestion is that, unless the physics community has reached consensus regarding a certain position, nobody else should speculate about or around it. The problem, of course, is that there is no consensus regarding *any* position when it comes to interpretations of QM, not only those the authors criticized. Therefore, if the authors were to have it their way, all popular debate regarding the implications of QM would cease.

I do not believe this outcome would be constructive. Although there is no consensus about what is the case, there is sufficient clarity and confidence about some very important things that—physicists already know—are *not* the case: naïve local-realism, for instance, has been categorically refuted, and this alone has tremendous repercussions in nearly every field of human activity. This is the elephant in the room. I do not believe that authors should close their eyes to it until physicists and philosophers have reached consensus about an alternative; I do not believe that popular writers in the fields of health care and parapsychology—to name only two—should pretend that business can continue as usual, as if naïve local-realism were true.

Although physicists are the authority when it comes to models of nature's behavior, they don't *own* their results. The discoveries of QM reveal the inner workings of nature and, as such, belong to *everyone*, for humans are all natural beings born from, and into, this universe. As such, people are all equally entitled—perhaps even morally required—to integrate these discoveries into their meditations about life, the universe, and everything; even—to the horror of the self-appointed skeptical police—those popular writers in the fields of health care and parapsychology.

Moreover, wild—and often ungrounded—speculation isn't a privilege of non-physicists. Today, physics itself is indulging in the most farfetched feast of speculations ever concocted by the human mind: multiple different types of parallel universes, each type potentially comprising a multidimensional infinity of such universes; ten spatial dimensions, many of which are supposedly curled up into tight little knots of extraordinary topological complexity; widely conflicting views about the nature of time, such as that time does not actually exist, that time is precisely the only thing that in fact exists (space being illusory), and that time exists but isn't fundamental, emerging instead from microscopic quantum processes; the accommodation of complete unknowns by mere labeling, such as the notions of dark matter and dark energy; widely differing views regarding the origin and early evolution of the universe; and the list goes on. Given all these seriously discussed hypotheses, it is difficult for physicists to take the moral high ground and criticize non-physicists based merely on the fact that the latter are engaging in physical speculation. Compared to the conjectures of many professional physicists today, allusions to quantum phenomena in health care and parapsychological literature sound rather moderate and conservative.

I acknowledge that this is not the authors' intended point or the spirit of their criticism. For them the problem is not *per se* that popular writers are engaging in physical speculation, but that these writers may be trying to misappropriate the authority of physics so to pass false or implausible claims for scientific fact. This kind of misappropriation is doubtlessly pernicious, dangerous, and I condemn it in the strongest terms.

However, as I have argued in this response, I do not believe that this really is what the popular writers the authors singled out were doing. What these writers stated in their works seems to me to be, by and large, reasonable enough—if poorly worded—given recent results from physics.

Although attempting to do something doubtlessly valid and important, I suspect the authors, by overshooting the mark, may have contributed to the very confusion they were trying to combat. This outcome is unfortunate, but it shouldn't stop efforts to separate wheat from chaff and bring some clarity to the reigning confusion around the foundations of quantum mechanics.

Chapter 19

Thinking Outside the Quantum Box

How mind can make sense of quantum physics in more ways than one

(The original version of this essay was published on Scientific American *on 16 February 2018)*

The counterintuitive predictions of quantum theory have now been experimentally confirmed with unprecedented rigor (Merali 2015). Yet, the question of how to interpret the meaning of these predictions remains controversial. At the time of this writing, a Wikipedia table summarizing different interpretations of quantum mechanics included no less than *fourteen* entries. New interpretations regularly appear.

The problem is that quantum theory contradicts our intuitive understanding of what 'real' means. According to the theory, if two real particles *A* and *B* are prepared in a special way, what Alice sees when she observes particle *A* depends on how Bob concurrently observes particle *B*, even if the particles—as well as Alice and Bob—are separated by an arbitrary distance. This "spooky action at a distance," as Einstein called it, contradicts either local causation or the very notion that particles *A* and *B* are 'real,' in the sense of existing independently of observation. As it turns out, certain statistical properties of the observations (Leggett 2003)—which have been experimentally confirmed (Cartwright 2007)—indicate the latter: that the particles do not exist independently of observation. And since observation ultimately consists of what is apprehended on the mental screen of perception, the implication may be that "the Universe is entirely mental," as bluntly put by Richard Conn Henry in a

Nature essay (2005).

The problem, of course, is that the hypothesis of a physical universe whose very existence depends on our minds contradicts mainstream scientific intuitions. So physicists scramble to interpret quantum theory in a way that makes room for a mind-independent physical reality. A popular way to do this entails postulating imagined, empirically unverifiable, theoretical entities *defined* as observer-independent. Naturally, this goes beyond mere interpretation: it adds redundant baggage to quantum theory, in the sense that the theory needs none of this stuff to successfully predict what it predicts.

Some cringe at such attempts to modify quantum mechanics to make it fit one's worldview, as opposed to adapting one's worldview to make it consistent with quantum mechanics. So the question that naturally arises is: If we stick to plain quantum theory, what does it tell us about reality? Physicist Carlo Rovelli tried to answer this question rigorously and the result is now known as relational quantum mechanics (RQM) (1996).

According to RQM, there are no absolute—that is, observer-independent—physical quantities. Instead, *all* physical quantities—the entire physical world—are relative to the observer, in a way analogous to motion. This is motivated by the fact that, according to quantum theory, different observers can account differently for the same sequence of events. Consequently, each observer is inferred to 'inhabit' its own physical world, as defined by the context of its own observations. And indeed, this astonishing prediction has been experimentally confirmed (Proietti *et al.* 2019, Proietti 2019, Emerging Technology from the arXiv 2019), a fact that has boosted RQM.

The price of this uncompromising honesty in acknowledging the implications of quantum mechanics is a number of philosophical qualms. First, the idea that the physical world one inhabits is a product of one's private observations seems to imply solipsism, an anathema in philosophy. Second, RQM entails that "a complete description of the world is exhausted by the relevant [Shannon] information that systems have about each other" (Rovelli 1996). However, according to Shannon (1948), information isn't a thing unto itself. Instead, it is constituted by the discernible configurations of a substrate. Yet, if there is no absolute physical substrate, what then constitutes information? Third—and perhaps most problematic of all—the RQM tenet that all physical quantities are relative raises an obvious question: *relative to what?* We only see meaning in a relative quantity such as motion because we assume there to be *absolute* physical bodies that move with respect to one another. But RQM denies *all* physical absolutes that could ground the meaning of relative quantities.

Notice that the root of all these philosophical qualms is the unexamined assumption that *only physical quantities exist*. If physical quantities arise from personal observation and they are all there is, then solipsism is indeed implied. If physical quantities are grounded in information and they are all there is, then information indeed lacks a substrate. If physical quantities are relative and they are all there is, then there are indeed no absolutes to ground their meaning. I shall return to this insight shortly.

For now, however, it would seem that biting the bullet of plain quantum theory, without decorating it with imagined bells and whistles, forces us into unresolvable philosophical qualms. Yet, such conclusion is false. To see how we can get out of this quagmire we need only to be rigorous about the epistemic scope of physics.

Stanford physicist Andrei Linde, of cosmic inflation fame, provided an important clue when he observed that

> our knowledge of the world begins not with matter but with perceptions ... Later we find out that our perceptions obey some laws, which can be most conveniently formulated if

> we assume that there is some underlying reality beyond our perceptions ... This assumption is almost as natural (and maybe as false) as our previous assumption that space is only a mathematical tool for the description of matter. (1998: 12)

Hence, in the absence of an absolute, observer-independent substrate, the physical world of RQM *can only be the contents of perception*. There is nothing else for it to be.

Now recall that the philosophical qualms of RQM rest on the assumption that only physical quantities—that is, contents of perception—exist. However—and here is the key point—*next to the contents of perception there are, of course, also non-perceptual mental categories such as thoughts*. Many physicists posit that thoughts should be explainable in terms of physical quantities and, as such, become part of the physical world by reduction. But this is a philosophical assumption that does not change the scientific fact that quantum mechanics does not predict thoughts; it only predicts the unfolding of perception, even when what is predicted—and later perceived—is the output of instrumentation.

So the possibility that presents itself to us is that thoughts are the absolutes that ground the meaning of the relative physical quantities of RQM. In other words, all physical quantities on the screen of perception may arise as relationships between thoughts. Moreover, since both thoughts and perceptions are mental in essence, this line of reasoning points to mind as the primary substrate of nature, the discernible states of which constitute information.

The hypothesis here, which I have elaborated upon in detail elsewhere (Kastrup 2019), is that thought—whose characteristic ambiguities may in fact be what quantum superposition states ultimately represent—underlies all nature and isn't restricted to living organisms. The physical world of an observing organism may arise from an interaction—an interference pattern—

between the organism's thoughts and the transpersonal thoughts underlying the inanimate universe that surrounds it. Although each organism—in accordance with RQM—may indeed inhabit its own private world of perceptions, all organisms may be surrounded by a common environment of thoughts, which avoids solipsism at least in spirit.

Conn Henry's courageous assertion that "the Universe is entirely mental" isn't only a seeming implication of recent experimental observations, it may also point the way to an elegant philosophical underpinning for what is perhaps the most rigorous and parsimonious interpretation of quantum mechanics. Mind, it seems, may offer a path out of the quantum quagmire in more ways than one.

Chapter 20

The Universe as Cosmic Dashboard

Relational quantum mechanics suggests physics might be a science of perceptions, not observer-independent reality

(The original version of this essay was published on Scientific American *on 24 May 2019)*

One of the weirdest theoretical implications of quantum mechanics is that different observers can give different—though equally valid—accounts of the same sequence of events. As highlighted by physicist Carlo Rovelli in his relational quantum mechanics (RQM) (1996), this means that there should be no absolute, observer-independent physical quantities. All physical quantities—the whole physical universe—must be relative to the observer. The notion that we all share the same physical environment must, therefore, be an illusion.

Such a counterintuitive prediction—which seems to flirt dangerously with solipsism—has been clamoring for experimental verification for decades. But only recently has technology advanced far enough to allow for it. So now, at last, Massimiliano Proietti and collaborators at Heriot-Watt University, in the UK, seem to have confirmed RQM: as predicted by quantum mechanics, there may well be no objective physical world (Proietti *et al.* 2019, Proietti 2019, Emerging Technology from the arXiv 2019).

Yet, our perceptions of the world beyond ourselves are quite consistent across observers: if you were to sit next to me right now, we would both describe my study in very similar, mutually consistent ways. Clearly, observers must share an environment of some sort, even if such an environment is not *physical*—i.e., not describable by physical quantities.

Possible solutions to this dilemma have been proposed. For instance, I maintain that physical quantities describe merely our *perceptions* and are, therefore, relative to each of us as observers (Kastrup 2019). What is really out there, underlying our perceptions, is constituted not by physical but by *transpersonal mental states* instead. Perceived physicality is merely a *cognitive representation* of that surrounding mental environment, brought into being by an act of observation.

This isn't a new view. In fact, it is very old. For instance, already in the early 19th century, Arthur Schopenhauer argued that the physical world of discrete objects in spacetime is merely a subjective representation in the mind of an observer (Kastrup 2020). What is really out there is what Schopenhauer called the "Will": transpersonal mental states with a volitional character, which transcend our ability to sense or measure directly. It is the volitional character of these states that explains the universe's evolution according to causal chains: the universe moves and changes because it is compelled to do so by the patterns of its own underlying willing.

Despite the objections one might have to Schopenhauer's ideas, they do seem to make sense of RQM's counterintuitive predictions: physics was developed to describe perceptual states alone, not endogenous mental states such as volition. For this reason, physical descriptions are always observer-dependent: they don't capture the world as it is in itself, but merely how it *presents* itself to each of us, given our respective point of view within the environment. Make no mistake: there still is a common environment of transpersonal volitional states, in which we are all immersed; it's just that this environment is not what physics directly describes.

Making sense of RQM by inferring that our surrounding environment is essentially mental—a view called 'objective idealism'—avoids solipsism. However, it carries with it a seemingly difficult problem: If what is really out there are transpersonal volitional states, then why do seeing or hearing feel so different from desiring or fearing? If my perceptions represent underlying states akin to desire and fear, why do I see forms and colors instead?

If only we could provide a compelling rationale for this qualitative transition, we would be able to leverage objective idealism to make sense of RQM and the latest experimental results. But can we? As it turns out, we very well can; even in more ways than one.

Over the past several years, Donald Hoffman's group at the University of California, Irvine, has shown that our perceptual apparatus hasn't evolved to represent the world truthfully, as it is in itself; if we saw the world as it really is, we would be swiftly driven to extinction (Hoffman 2009, Hoffman & Singh 2012). Instead, we see the world in a way that favors our survival, not the accuracy of our representations. In Hoffman's analogy, the contents of perception are like icons on a computer desktop: a set of visual metaphors that facilitate one's job by illustrating the salient properties of files and applications, but which don't portray these files and applications as they really are.

Approaching the problem from a different angle, Karl Friston and collaborators have shown that, if an organism is to represent the states of the external environment in order to properly navigate this environment, it must do so in an *encoded, inferential manner* (Friston, Sengupta & Auletta 2014). The reason is that, if the organism were to simply mirror the states of the external environment in its own internal states, it would not be able to maintain its structural integrity: its internal states would become too dispersed and the organism would dissolve into an entropic soup. Perceptual encoding is necessary for the organism to resist entropy and thus remain alive.

What both of these lines of argument suggest is this: the

screen of perception is much more akin to a *dashboard* than a window into the environment. It conveys relevant information about the environment in an indirect, encoded manner that helps us survive. The forms and colors we see, the sounds we hear, the flavors we taste are all like *dials*: they present to us, at a glance, information that correlates—in a manner fundamentally beyond our ability to cognize—with the mental states of the environment out there.

Instead of having to directly *feel* the myriad mental states surrounding us—which would be akin to how a telepath would feel overwhelmed and disoriented in the middle of an agitated crowd—we encode them neatly in the pixels of the screen of perception. Or—to use another metaphor—instead of airplane pilots who must look through the windshield in the middle of an electric storm, we fly by instruments.

Evolution has provided each of us with a dashboard of dials that inform us about the environment we live in. But we don't have a window to look directly at what is out there; all we have are the dials. *The error we make is in mistaking the dials for the external environment itself.*

Physics models and predicts the behavior of the dials. Although we are all immersed in a common environment, each of us interacts with it in a different way, from a different perspective. Therefore, we each gather different information about the environment, and so our respective dials may not always agree. This doesn't mean that there is no common environment; it means only that this environment isn't *physical*.

For as long as we insist that the world, *as it is in itself*, must have the forms and contours of the images on the screen of perception, quantum mechanics will continue to be paradoxical. For as long as we believe that physical theory models the shared environment underlying our perceptions—as opposed to the *perceptions themselves*—quantum mechanics will continue to be puzzling. There is only one reasonable way out: to regard our perceptions as a dashboard of dials providing salient, though indirect, information about a mental universe out there.

Chapter 21

Physics Is Pointing Inexorably to Mind

The reasoning behind so-called 'information realism' has some unexpected implications

(The original version of this essay was published on Scientific American *on 25 March 2019)*

In his book, *Our Mathematical Universe,* physicist Max Tegmark boldly claims that "protons, atoms, molecules, cells and stars" are all redundant "baggage" (2014: 255). Only the mathematical apparatus used to *describe* the behavior of matter is supposedly real, not matter itself. For Tegmark, the universe is a "set of abstract entities with relations between them," which "can be described in a baggage-independent way" (*Ibid.*: 267)—i.e., *without* matter. He attributes existence solely to descriptions, while incongruously denying the very thing that is described in the first place. Matter is done away with and only information itself is taken to be ultimately real.

This abstract notion—called 'information realism'—is intrinsically philosophical in character, but it has been associated with physics from its very inception. Most famously, information realism is a popular philosophical underpinning for 'digital physics,' a set of theoretical perspectives according to which the universe is the output of something akin to a cosmic computer program (Fredkin 2003).

Indeed, according to the Greek atomists, if we kept on dividing things into ever-smaller bits, at the end there would remain solid, indivisible corpuscles called 'atoms,' imagined to be so concrete as to even have particular shapes. Yet, as our understanding of physics progressed, we've realized that atoms themselves can be further divided into smaller bits, and those into yet smaller ones, until what is left lacks shape and solidity altogether. At the bottom of the chain of physical reduction there are only elusive, phantasmal entities we label as 'energy' and 'fields': abstract conceptual tools for describing nature, but which themselves seem to lack any real, concrete essence.

To some physicists, this indicates that what we call 'matter'—with its solidity and concreteness—is an illusion; that only the mathematical apparatus they devise in their theories is truly real, not the perceived world the apparatus was created to describe in the first place. From their point of view, such a counterintuitive conclusion is an implication of theory, not a conspicuously narcissistic and self-defeating proposition.

Indeed, according to information realists, matter arises from information processing, not the other way around. Even mind—psyche, soul—is supposedly a derivative phenomenon of purely abstract information manipulation (Fredkin n.a.). But in such a case, what exactly is meant by the word 'information,' since there is no physical or mental substrate to ground it?

You see, it is one thing to state in language that information is fundamental and can, therefore, exist independently of mind and matter. But it is another thing entirely to *explicitly* and *coherently* conceive of what—if anything--this may mean. By way of analogy, it is possible to write—as Lewis Carroll did—that the Cheshire Cat's grin remains after the cat disappears, but it is another thing altogether to conceive explicitly and coherently of what this means.

Our intuitive understanding of the concept of information—as cogently captured by Claude Shannon (1948)—is that it is merely a measure of the number of possible states an independently existing system can assume. As such, information is a property *of an underlying substrate*—associated with the substrate's possible configurations—not an entity unto itself.

To say that information exists in and of itself is akin to

speaking of spin without the top, of ripples without water, of a dance without the dancer, or of the Cheshire Cat's grin without the cat. It is a grammatically valid statement devoid of sense; a word game less meaningful than fantasy, for internally consistent fantasy can at least be explicitly and coherently conceived of as such.

One assumes that serious proponents of information realism are well aware of this line of criticism. How do they then reconcile their position with it? A passage by Luciano Floridi may provide a clue. In a section titled "The nature of information," he states:

> Information is notoriously a polymorphic phenomenon and a polysemantic concept so, as an explicandum, it can be associated with several explanations, depending on the level of abstraction adopted and the cluster of requirements and desiderata orientating a theory. ... *Information remains an elusive concept.* (2008: 117, emphasis added)

Such obscure ambiguity lends information realism a conceptual fluidity that renders it unfalsifiable. After all, if the choice of primitive is given by "an elusive concept," how can one definitely establish that it is wrong? In admitting the possibility that information may be "a network of logically interdependent but mutually irreducible concepts" (*Ibid.*: 120), Floridi seems to suggest even that such elusiveness is inherent and unresolvable.

Whereas vagueness may be defensible in regard to natural entities conceivably beyond the human ability to fathom, it is difficult to justify when it comes to a *human* concept, such as information. *We invented the concept*, so we either specify clearly what we mean by it or our conceptualization remains too vague to be meaningful. In the latter case, there is literally *no sense* in attributing primary existence to information.

The untenability of information realism, however, does not erase the problem that motivated it to begin with: the realization that, at bottom, what we call 'matter' becomes pure abstraction, a phantasm. How can the felt concreteness and solidity of the perceived world evaporate out of existence when we look closely at matter?

To make sense of this conundrum, we don't need the word games of information realism. Instead, we must stick to what is most immediately present to us: solidity and concreteness are *qualities of our experience*. The world measured, modeled and ultimately predicted by physics is the world of *perceptions*, a category of mentation. The phantasms and abstractions reside merely in our *descriptions* of the behavior of that world, *not in the world itself.*

Where we get lost and confused is in imagining that what we are describing is a non-mental reality underlying our perceptions, as opposed to *the perceptions themselves*. We then try to find the solidity and concreteness of the perceived world in that postulated underlying reality. However, a non-mental world is inevitably abstract. And since solidity and concreteness are *felt qualities of experience*—what else?—we cannot find them there. The problem we face is thus merely an artifact of thought, something we conjure up out of thin air because of our theoretical habits and prejudices.

Tegmark is correct in considering matter—defined as something outside and independent of mind—to be unnecessary baggage. But the implication of this fine and indeed brave conclusion is that *the universe is a mental construct* displayed on the screen of perception. Tegmark's "mathematical universe" is inherently a *mental* one, for where does mathematics—numbers, sets, equations—exist if not in mentation?

None of this implies solipsism. The mental universe exists in mind but *not in your personal mind alone*. Instead, it is a *trans*personal field of mentation that presents itself to us as physicality—with its concreteness, solidity and definiteness—once our personal mental processes interact with it through observation. This mental universe is what physics is leading us to, not the hand-waving word games of information realism.

An afterthought

Following the publication of the original version of this essay on *Scientific American*, a number of other media outlets picked up the story. Incomprehensibly, however, they misportrayed it as—of all things—an argument *in favor of information realism* (!), proceeding to promote such an abysmal misunderstanding with all the furor that usually accompanies counterintuitive claims originating from respectable sources.

That an essay whose central argument *criticizes* and *debunks* information realism is misportrayed as *promoting* it speaks volumes to the laziness, gullibility, sensationalism, lack of professionalism and utter unreliability that—tragically—reigns in some segments of the media. I sometimes despair not only at having to witness my words being distorted to the point that their meaning is inverted; not only at the incalculable damage this does to the credibility of the media; but also—and especially—at the lamentable situation that the average educated person finds themselves in when trying to make sense of the world in the age of social media.

Chapter 22

Do We Actually Experience the Flow of Time?

Subjective experience must inform physics and philosophy, but it should be assessed carefully

(The original version of this essay was published on Scientific American *on 14 November 2018)*

Time is a contentious topic in physics. Some physicists, such as Julian Barbour, argue that it doesn't even exist (1999). Others, such as Carlo Rovelli, hold that it arises as a secondary effect of deeper quantum processes (2018). Yet others, such as Lee Smolin, maintain that time is the sole fundamental dimension of nature (2013). And because the laws of physics are time-symmetrical, much debate has gone into figuring out why we seem unable to travel back in time.

All this theorizing is motivated by—and attempts to make sense of—our subjective experience of the forward flow of time. Indeed, our reliance on what we think we experience as the flow of time goes so deep that some philosophers take it for a self-evident axiom. For instance, Susan Schneider has claimed that the flow of time is inherent to experience—so much so that, according to her, "timeless experience is an oxymoron" (2018).

But do we *actually* experience the flow of time? We certainly experience something that looks like it. But if we introspect carefully into this experience, is what we find accurately describable as 'flow'?

There can only be experiential flow if there is experience in the past, present and future. But where is the past? Is it anywhere out there? Can you point at it? Clearly not. What makes you conceive of the idea of the past is the fact that you have memories. But these memories can only be referenced insofar as they are experienced *now*, as memories. There has never been a single point in your entire life in which the past has been anything other than memories experienced in the present.

The same applies to the future: Where is it? Can you point at it and say, "There is the future"? Clearly not. Our conception of the future arises from expectations or imaginings experienced *now*, always now, as expectations or imaginings. There has never been a single point in your life in which the future has been anything other than expectations or imaginings experienced in the present.

But if the past and the future are not actually experienced in the ... well, past and future, how can there be an experiential *flow* of time? Where is experiential time flowing from and into?

Let's make an analogy with space. Suppose that you suddenly find yourself sitting on the side of a long, straight desert road. Looking ahead, you see mountains in the distance. Looking behind, you see a dry valley. The mountains and the valley provide references that allow you to locate yourself in space. But the mountains, the valley, your sitting on the roadside, all exist simultaneously in the present snapshot of your conscious life.

An entirely analogous situation occurs in time: right now, you find yourself reading this essay. As you read it, you can remember having done something else—say, having brushed your teeth—earlier today. You can also imagine that you will do something else later—say, lie down in bed. Brushing your teeth and lying down in bed are respectively behind and ahead of you on the road of time—your 'timescape'—just as the valley and the mountains were on the road of space. They provide references that allow you to locate yourself in time. But again, the experiences of remembering the past and imagining the future, as well as that of reading this essay right now, all exist simultaneously in the *present* snapshot of your conscious life.

The problem is that we then construe from this that there is an experiential flow of time. Such a conclusion is as unjustifiable as to construe, purely from seeing the mountains ahead and the valley behind while you sit by the roadside, that you are moving on the road. You aren't; you are simply taking account of your relative position on it. You have no more experiential reason to believe that time flows than that space flows while you sit quietly by the roadside.

You may claim that, whereas the desert road scenario is static, lacking action, you actually *did* brush your teeth earlier. So time definitely flowed from then to now; or did it? All you have to support the belief that it did is your memory of having brushed your teeth, which you experience now. All you ever have is the present experiential snapshot. Even the notion of a previous or subsequent snapshot is—insofar as you can know from experience—merely a memory or expectation *within* the present snapshot. The flow from snapshot to snapshot is a story you tell yourself, irresistibly compelling as it may be. Neuroscience itself suggests that this flow is indeed a cognitive construct (Buonomano 2018, Eagleman 2009).

A thought experiment may help: suppose that you could return to your past—say, back to the moment when you were brushing your teeth this morning. In the corresponding experiential snapshot, the present would lie between, say, the memory of your having stood up from bed and the expectation of your dressing up for work. But once you landed on that snapshot, you would have no experience of any temporal discontinuity: you would look behind in memory and see yourself standing up from bed; you would look ahead in imagination and see yourself dressing up for work. The tape of history would have been rewound and you would have no memory of having time-traveled; otherwise you wouldn't have *actually* time-traveled. Everything would feel perfectly normal—just as it feels right now. So who is to say that you haven't time-traveled a moment ago? How do you know that time always flows forward?

You see, whether time flows forward, or doesn't flow at all, or moves back and forth, our resulting subjective experience would be identical in all cases: we would always find ourselves in an experiential snapshot extending smoothly backwards in memory and forwards in expectation, just like the desert road. We would always tell ourselves the same story about what's going on. A mere cognitive narrative—based purely on contents of the experiential snapshot in question—would suffice to convince us of the forward flow of time even when such is not the case.

The ostensible experience of temporal flow is thus an illusion. All we ever actually experience is the present snapshot, which entails a timescape of memories and imaginings analogous to the landscape of valley and mountains. Everything else is a story. The implications of this realization for physics and philosophy are profound. Indeed, the relationship between time, experience and the nature of reality is liable to be very different from what we currently assume. To advance our understanding of reality we must thus revise cherished assumptions about our experience of time.

Chapter 23

Why Does Nature Mirror Our Reasoning?

What the bizarre connections between mathematics and physics are telling us

A recent paper (Ji, Z. *et al.* 2020) is bringing back to light an old mystery that unites physics and mathematics in surprising ways. The paper shows that quantum entanglement—a concrete physical phenomenon that Einstein famously described as "spooky action at a distance"—can help mathematicians gain insight into very complex but strictly abstract mathematical conjectures.

This isn't the first time that an unexpected bridge linking physics to abstract mathematics has been uncovered. A few years ago, Oxford mathematician Minhyong Kim found a way to tackle problems in abstract number theory by visualizing the solutions as trajectories of light. "If the connection sounds fantastical it's because it is, even to mathematicians," wrote Kevin Hartnett at the time (2017).

These mysterious connections work the other way around as well: in an address to the Prussian Academy of Sciences, Albert Einstein asked, "How can it be that mathematics, being after all a product of human thought which is independent of [empirical] experience, is so admirably appropriate to the objects of reality?" (1921) Perhaps most famously, in 1960 Nobel laureate physicist Eugene Wigner published a paper titled "The Unreasonable Effectiveness of Mathematics in the Natural Sciences." The essence of the paper was a discussion about "the miracle of the appropriateness of the language of mathematics for the formulation of the laws of physics."

To illustrate his point, Wigner recounted an anecdote: a

professor of statistics showed to a friend some graphs and equations that modeled human population trends. In one of the equations, the friend noticed the symbol 'π' (pi) and asked what it was. The professor then explained that 'π' is the number you get when you divide the circumference of the circle by its diameter. The friend reacted with incredulity: "surely the population has nothing to do with the circumference of the circle" (Wigner 1960). Yet, somehow it does.

The sheer number of bizarre connections between purely abstract mathematics and concrete physics is remarkable: an imaginary vector space devised by mathematician David Hilbert turns out to represent physical quantum states; an imaginary notion of curvature developed by mathematician Bernhard Riemann turns out to describe the hidden geometry of spacetime according to general relativity; physicist Paul Dirac's equation to model the behavior of the electron contained a term with no known physical correspondence at the time, which turned out to predict the existence of the positron (duly found a few years later); and so on.

To this day, much of physics is guided by hints and predictions derived from the purely mental world of mathematics. Indeed, some physicists—such as Max Tegmark—have gone as far as to say that the physical world *is* mathematics, matter being redundant conceptual baggage (2014). This, in my view, overshoots the mark in that it tries to replace a description for that which is described in the first place; but it does illustrate the amazing connection between *mental* mathematics and *physical* world.

The obvious question then is: Why should the physical world of tables and chairs mirror the patterns of the mental world of abstract mathematical reasoning? Why should conclusions from pure thought match empirical observations? This is difficult to make sense of, which is probably why Wigner used the word 'miracle' twelve times in his famous paper.

Short of an extraordinarily implausible fluke, the connections suggest that the mental world of mathematical reasoning is somehow *continuous with* the physical world of tables and chairs. Under the hood, these seemingly different worlds must somehow be essentially the same: either the mental world of mathematics is essentially physical, or the ostensibly physical world of tables and chairs is essentially mental.

Commonsense as it may be, the first option fails as an explanation for Wigner's miracle: while the abstract mathematical apparatus in question—Riemannian geometry, Hilbert spaces, field equations, etc.—doubtlessly governs the physical structure and dynamics of the brain, it would also need to be *mirrored by the thoughts* the brain produces; otherwise we wouldn't be aware of it.

Now, why would thoughts mirror the physical intricacies of the brain? Bile doesn't mirror the anatomy or physiology of the liver; insulin doesn't mirror the anatomy or physiology of the pancreas; why should it be different when it comes to thoughts? Moreover, we don't need to know how the engine of a car works so as to effectively drive the car; we don't need to know how the electronic chips in a computer work so as to effectively use the computer; why would we need to know how the brain works—at the level of elementary subatomic particles—so as to survive and reproduce? No, the physical characteristics of the brain cannot account for Wigner's miracle.

But if the physical world—despite appearances to the contrary—is *mental* in essence, the connections between mathematics and physics become perfectly understandable: the basic, fundamental thought-templates that organize our mathematical reasoning *also organize the world itself*. Therefore, it is only natural that objective physical phenomena should comport themselves in a way consistent with our subjective mathematical logic: both have the same root, obey the same organizing principles. The apparent eeriness of Wigner's miracle melts away.

To postulate that the physical world is mental in essence does not contradict, or require changes in, physics. The idea is merely that what we call 'matter' is the extrinsic appearance of inner mental activity, just as our brain is what our thoughts look like when observed from the outside. Physics describes these extrinsic appearances—by means of abstract quantities and corresponding mathematical equations—irrespective of what their inner essence is. Indeed, we've known at least since the early 20th century—thanks to the work of Bertrand Russell (2009)—that physics is agnostic of the inner essence of things.

The mental universe hypothesis is thus entirely compatible with physics and, by implication, with all of science as we know it. As a matter of fact, it is arguably the only hypothesis that remains consistent with the latest experimental results emerging from quantum mechanics (see Chapters 16, 17, 20 and 21 of this book), as well as some peculiar observations in neuroscience (see Chapter 25). Finally, it is a hypothesis that has proven to be coherent after extensive theoretical treatment (Kastrup 2019). That it is also the most plausible hypothesis for making sense of Wigner's miracle is the cherry on the cake.

The future of physics—and of all of science—is mind; not your or my individual mind alone, but mind as a transpersonal essence that gives matter its inner reality. Therein lie the solutions to the conundrums of quantum mechanics, the mind-body problem and the otherwise bizarre connections between mathematics and physics.

Chapter 24

Is Life More Than Physics?

Our existence may be governed by yet-unrecognized organizing principles

A working hypothesis in science is that almost everything in the universe can be explained in terms of simpler parts. The simplest of these parts are the elementary subatomic particles, fundamental building blocks of nature. We assume that we could, in principle, reconstruct everything else in the universe by assembling these basic building blocks together in just the right way, like Lego bricks. Nobel laureate physicist Philip Anderson called this the "constructionist hypothesis" (1972).

The idea behind it is that, given the equations that govern the behavior of the subatomic particles, we should be able to predict all natural phenomena, including the largest and most complex. If so, understanding the physical laws that operate at the microscopic level—the level of the subatomic particles—is sufficient to provide us with a 'Theory of Everything.'

In practice, however, it is impossible to predict the behavior of all but the simplest and most minuscule natural phenomena based on this theory. In the words of Nobel laureate physicist Robert Laughlin and David Pines, the associated equations

> cannot be solved accurately when the number of particles exceeds about 10. No computer existing, *or that will ever exist,* can break this barrier because it is a catastrophe of dimension. ... Predicting protein functionality or the behavior of the human brain from these equations is patently absurd. (2000, emphasis added)

Laughlin and Pine's point is that, each time we add a particle to the system we are trying to model, its complexity grows exponentially. And exponential growth—such as in Ponzi schemes and nuclear chain reactions, to mention two radically different examples—becomes unmanageable very quickly. It is thus effectively impossible to use the Theory of Everything to model relevant, real-life systems. In Laughlin and Pine's words,

> We have succeeded in reducing all of ordinary physical behavior to a simple, correct Theory of Everything only to discover that it has revealed exactly nothing about many things of great importance. (Ibid.)

Now, since we cannot test the Theory of Everything in real life, we just don't know whether it is sufficient to explain ... well, *real life*; for even the simplest metabolizing organisms comprise bazillions of particles.

Aside from practical limitations, there are also theoretical and empirical reasons to suspect that fundamental organizing principles exist in nature at a macroscopic level—that is, at the level of things we can touch and manipulate with our bare hands and see with naked eyes. If so, these macroscopic principles aren't captured by the Theory of Everything. Instead, they are extra, yet-unrecognized natural laws.

Philip Anderson himself posited this hypothesis. He alluded to life as a phenomenon we may never be able to explain in terms of subatomic particles, because it may be governed by yet-unrecognized natural laws operating at a macroscopic level:

> we have yet to recover from [the arrogance] of some molecular biologists, who seem determined to try to reduce everything about the human organism to 'only' chemistry ... [However,] each level [of biological organization] can require a whole new conceptual structure. (1972)

Anderson's views were echoed in 2008 by Mile Gu and collaborators:

> complex systems may possess emergent properties difficult or impossible to deduce from a microscopic picture ... macroscopic laws that govern macroscopic observables ... cannot logically be derived, *even in principle*, from microscopic principles. (emphasis added)

Debated as it still is, such hypothesis is acknowledged even by orthodox physicists such as Sabine Hossenfelder, who granted that "a derivation of emergent from fundamental properties [might turn out to be] impossible even theoretically" (2010). Where does this leave us? If there are fundamental, macroscopic organizing principles governing life—that is, natural laws that only kick in at the level of large organic molecules and living tissue—why haven't we found them yet?

To see why, notice first that we find a law of nature by observing—under controlled conditions—a regularity in nature's behavior. Repeated observations of the same behavior under the same circumstances (such as a stone dropping to the ground every time it is released from one's hand) motivate us to infer a corresponding law (such as gravity). The observations of microscopic behavior underpinning the Theory of Everything are possible because scientists shield the experimental setup from the environment, so as to control its conditions.

However, if there are natural laws that kick in only at the level of living organisms, it is effectively impossible to isolate and pin them down: metabolizing organisms are just too complex to experiment with under controlled conditions at the level of irreducibility. Moreover, it would be exceedingly difficult to even *notice* these hypothetical laws in action in the first place: the complexity of metabolism prevents us from keeping tabs on all the salient chains of cause and effect. Instead, we would be tempted to simply *assume*—as we in fact do—that everything can somehow be traced back to the behavior of subatomic particles, according to the Theory of Everything.

There are thus multiple avenues of argument converging towards the possibility that the natural laws governing our bodies and actions go beyond the recognized laws of microscopic physics. The notion that everything about life can be reduced to the behavior of subatomic particles is just a guess; we just aren't in a position to know.

And so, we may be forced to reconsider some neglected hypotheses raised by scientists and philosophers of the past. For instance, in the early 19th century Arthur Schopenhauer hypothesized that living organisms are governed by more than just the physical and chemical laws that rule the inorganic world:

> we shall certainly find in the organism traces of chemical and physical modes of operation, but we shall never explain the organism from these, because it is [brought about by] a higher Idea that has subdued these lower ones (as quoted in Kastrup 2020).

In other words, for Schopenhauer life is created and maintained by a "higher" organizing principle operating at the level of the organism *as a whole*, not just its constituent atoms and molecules. And although we have advanced immensely since the early 19th century, today we still can't refute Schopenhauer's hypothesis.

Neither can we refute the synchronicity hypothesis raised in the 20th century by psychiatrist Carl Jung and Nobel laureate physicist Wolfgang Pauli—of the 'Pauli exclusion principle'—in their extensive correspondence (Jung & Pauli 2001): in addition to the known microphysical laws, Jung and Pauli speculated that the events of life may be orchestrated by a macroscopic ordering principle that tends to bring similar events together (cf. Kastrup 2021). According to this hypothesis, instead of mere flukes, some meaningful coincidences are the natural outcome of such orchestration.

There may be much more to life than we dream of today; unsuspected organizing principles in nature that influence—perhaps even govern—how our bodies work, what we think and feel, and how we act in the world. Investigating this possibility scientifically is perhaps one of the greatest and most important challenges yet to be addressed, for it directly concerns our own essential nature as living beings.

Part VI

On Psychology and Neuroscience

Chapter 25

Transcending the Brain

At least some cases of physical brain damage are associated with enriched consciousness or cognitive skill

(The original version of this essay was published on Scientific American *on 29 March 2017)*

Despite significant advances in neuroscience, consciousness remains a vexing mystery. Because the qualities of experience seem to be irreducible to physical parameters (Chalmers 2003), a hypothesis that has been garnering attention is that consciousness is fundamental and spatially unbound, the brain corresponding to a dissociation or localization of its contents (Kastrup 2017e, 2018a, Shani 2015, Nagasawa & Wager 2016). At first sight, simple observation seems to contradict this hypothesis: as pointed out by neuroscientist Sam Harris, if a normally functioning brain corresponds to a limitation of cognition,

> ... one would expect most forms of brain damage to unmask extraordinary scientific, artistic, and spiritual insights. ... A few hammer blows or a well-placed bullet should render a person of even the shallowest intellect a spiritual genius. Is this the world we are living in? (2012b)

Harris's rhetorical question alludes to the indisputable fact that most forms of brain function impairment correlate with cognitive deficit. However, a more interesting question is perhaps this: Do *some* forms of impairment correlate with an enrichment of consciousness or cognitive skill? After all, even if only one black swan can be conclusively discerned in a herd of white swans, our theories about the origin and nature of swans must be able to make sense of those black individuals.

As it turns out, there are reliable reports in the medical literature of—yes—bullet wounds to the head, stroke, concussion, meningitis and even the progression of dementia leading to expanded cognitive and artistic skills (Lythgoe *et al.* 2005, Treffert 2006, 2009: 1354, Piore 2013, Miller *et al.* 1998, 2000). Ironically, therefore, Harris's rhetorical question has an affirmative answer: somehow, this is indeed the world we are living in.

And that's just the tip of the iceberg. Many forms of brain function impairment associated with seeming unconsciousness are now known to be accompanied by richer inner life. For instance, the dangerous 'choking game' played by teenagers worldwide (Macnab 2009) is an attempt to induce rich feelings of self-transcendence through partial strangulation and fainting (Neal 2008: 310-315). The psychotherapeutic technique of holotropic breathwork (Rhinewine & Williams 2007) also uses hyperventilation-induced fainting to achieve what is described as an expansion of awareness (Taylor 1994). Even pilots undergoing 'G-force induced Loss Of Consciousness' (G-LOC)—whereby blood is forced out of the brain—report "memorable dreams" (Whinnery & Whinnery 1990).

Generalized physiological stress caused, for instance, by cardiac arrest—which severely compromises brain function—is sometimes accompanied by reports of 'Near-Death Experiences' (NDEs) (van Lommel 2001). NDEs reportedly entail life-transforming insights, emotions and inner imagery far richer than ordinary experiences (Kelly *et al.* 2009: 367-421), despite overwhelming disruption to the brain's ability to operate.

This pattern of correlations between brain function impairment and a seeming expansion of awareness is surprisingly broad. For instance, during the practice of so-called 'psychography,' an alleged medium enters a trance state and writes down information allegedly originating from a transcendent source. A detailed neuroimaging study revealed that experienced mediums displayed marked reduction of activity in key brain regions—such as the frontal lobes and hippocampus—when compared to regular, non-trance writing (Peres 2012). Despite this, text written under trance scored consistently higher in a measure of complexity than material produced without trance.

Even more intriguingly, it is well known that psychedelic substances induce powerful experiences of self-transcendence (Strassman 2001, Griffiths *et al.* 2006, Strassman *et al.* 2008). It had been assumed that they did so by exciting parts of the brain. Yet, recent neuroimaging studies have shown that psychedelics do largely the opposite (Carhart-Harris *et al.* 2012, 2016, Muthukumaraswamy *et al.* 2013, Palhano-Fontes *et al.* 2015, Lewis *et al.* 2017). Moreover, "the magnitude of this decrease [in brain activity] predicted the intensity of the subjective effects" (Carhart-Harris *et al.* 2012: 2138). In other words, the *less* activated the brain becomes, the *more* intense the psychedelic experiences are.

If this pattern is consistent, we should expect some types of physical brain damage to also lead to experiences of self-transcendence. And indeed, this has been reported. In a relatively recent study, CT scans of more than one hundred Vietnam war veterans showed that damage to the frontal and parietal lobes increased the likelihood of "mystical experiences" (Cristofori 2016). In an earlier study, patients were evaluated before and after brain surgery for the removal of tumors, which caused collateral damage to surrounding tissue. Statistically significant increases in "feelings of self-transcendence" were reported after the surgery (Urgesi *et al.* 2010).

Clearly, there is a broad and consistent pattern associating impairment of brain function with—in the words of Harris (2012b)—"extraordinary scientific, artistic, and spiritual insights." That this happens in but a small minority of cases isn't surprising: damage affecting memory pathways, metacognition, language centers, or any other cognitive function necessary for recalling or reporting inner life erases the signs of such insights. A person lying in a vegetative state could be having indescribably rich inner experiences and we would be none the wiser. The evidence is necessarily constrained to a narrow window between brain function impairment insufficient to trigger self-transcendence and impairment that renders self-transcendence unreportable to self or others.

It is conceivable that brain function impairment could disproportionally affect inhibitory neural processes, thereby generating or bringing into awareness other neural processes associated with self-transcendence. However, if experience is constituted, generated, or at least fully modulated by brain activity, an increase in the richness of experience must be accompanied by an increase in the metabolism associated with the neural correlates of experience (Kastrup 2016b). Any other alternative would decouple experience from the workings of the living brain information-wise. As such, it is difficult to see how partial strangulation, hyperventilation, G-LOC, cardiac arrest, etc.—which reduce oxygen supply to the brain as a whole—could selectively affect inhibitory neural processes whilst preserving enough oxygen supply to fuel an increase in the neural correlates of experience.

Alternatively, one could speculate that experiences of self-transcendence occur only after normal brain function resumes. This, however, cannot account for several of the cases mentioned. For instance, during the neuroimaging studies of the psychedelic state researchers collected subjective reports of self-transcendence while *concurrently* monitoring the subjects' reduced brain activity levels. The same holds for the neuroimaging study of psychography. Finally, in cases of acquired savant syndrome the savant skills are often concomitant with the presence of physical damage in the brain.

It is conceivable that individual cases of self-transcendence could have their own idiosyncratic explanation, unrelated to the other cases, and that the overall pattern suggested here is a red herring. However, all cases mentioned here, besides being associated with brain function impairment, also share strikingly consistent subjective reports. Consider the two passages below:

Passage 1:

> I certainly don't feel reduced or smaller in any way. On the contrary, I haven't ever been this huge, this powerful, or this all-encompassing. ... [I] felt greater and more intense and expansive than my physical being. (Moorjani 2012: 69)

Passage 2:

> My perception of my physical boundaries was no longer limited to where my skin met air. I felt like a genie liberated from its bottle. The energy of my spirit seemed to flow like a great whale gliding through a sea of silent euphoria. (Taylor 2009: 67)

Passage 1 was reported by the subject of an NDE caused by generalized physiological stress, while passage 2 was reported by the subject of a stroke.

Such similarities suggest that normal brain function corresponds to a dissociation or localization of the contents of consciousness, and that certain forms of *impairment* of brain function reduce this dissociation or localization, thereby leading to *expanded* awareness and self-transcendence. The implications of this hypothesis for both neuroscience and neurophilosophy are far-reaching.

Chapter 26

Consciousness Goes Deeper Than You Think

Metacognitive awareness is part of it, but it's much more than that

(The original version of this essay was published on Scientific American *on 19 September 2017)*

An article on the neuroscience of infant consciousness, which attracted some interest a few years ago, asked: "When does your baby become conscious?" (Gabrielsen 2013). The premise, of course, was that babies aren't born conscious but, instead, develop consciousness at some point (according to the article, at about five months of age). Yet, it is hard to think that there is nothing it feels like to *be* a newborn.

Newborns clearly seem to experience their own bodies, environment, the presence of their parents, etc.—albeit in an unreflective, present-oriented manner (Koch 2009). And if it always feels like something to *be* a baby, then babies don't *become* conscious; instead, they are conscious from the get-go.

The problem is that—somewhat alarmingly—the word "consciousness" is often used in the literature as if it entailed or implied more than just the qualities of experience. Dijksterhuis and Nordgren, for instance, insisted that "it is very important to realize that attention is the key to distinguish[ing] between unconscious thought and conscious thought. Conscious thought is thought with attention" (2006). This implies that if a thought escapes attention, then it is unconscious. But is the mere lack of attention enough to assert that a mental process lacks the qualities of experience? Couldn't a process that escapes the focus of attention still feel like something?

Consider your breathing right now: the sensation of air flowing through your nostrils, the movements of your diaphragm, etc. Were you not experiencing these sensations a moment ago, before I directed your attention to them? Or were you just unaware *that* you were experiencing them all along? By directing your attention to these sensations, did I make them conscious or did I simply cause you to experience the extra quality of knowing *that* the sensations were conscious?

Indeed, Jonathan Schooler has established a clear distinction between conscious and *meta*-conscious processes (2002). Whereas both types entail the qualities of experience, meta-conscious processes also entail what he called *re-representation*:

> Periodically attention is directed towards explicitly assessing the contents of experience. The resulting meta-consciousness involves an explicit re-representation of consciousness in which one interprets, describes or otherwise characterizes the state of one's mind. (Ibid.)

So where attention plays an important role is in re-representation; that is, the conscious knowledge *of* an experience, which underlies introspection. Subjects cannot report—not even to themselves—experiences that aren't re-represented. Nothing, however, stops conscious experience from occurring without re-representation: dreams, for instance, have been shown to largely lack re-representation (Windt & Metzinger 2007), despite the undeniable fact they are experienced in consciousness. Typically, we only re-represent the *memory* of the dream—upon awakening—not the dream itself, while it is occurring. This gap between reportability and the contents of consciousness has motivated the emergence of so-called 'no-report paradigms' in the modern neuroscience of consciousness (Tsuchiya *et al.* 2015).

Clearly, the assumption that consciousness is limited to

re-represented mental contents under the focus of attention mistakenly conflates meta-consciousness with consciousness proper. Yet, this conflation is disturbingly widespread. Consider Axel Cleeremans' words:

> Awareness ... always seems to minimally entail the ability of knowing *that* one knows. This ability, after all, forms the basis for the *verbal reports* we take to be the most direct indication of awareness. And when we observe the absence of such *ability to report* on the *knowledge* involved in our decisions, we rightfully conclude the decision was based on unconscious knowledge. (2011, emphasis added)

Because the study of the Neural Correlates of Consciousness (NCC) is, by and large, dependent on subjective reports of experience (Newell & Shanks 2014), what passes for the NCC is liable to be merely the neural correlates of *meta*-consciousness. As such, potentially conscious mental activity—in the sense of activity correlated with experiential qualities—may evade recognition as such.

As a matter of fact, there is circumstantial but compelling evidence that this is precisely the case. To see it, notice first that the conscious knowledge *N*—that is, the re-representation—of an experience *X* is triggered by the occurrence of *X*. For instance, it is the occurrence of a sense perception that triggers the metacognitive realization that one is perceiving something. *N*, in turn, evokes *X* by directing attention back to it: the realization that one is perceiving something naturally shifts one's mental focus back to the original perception. So we end up with a back-and-forth cycle of evocations whereby *X* triggers *N*, which in turn evokes *X*, which again triggers *N*, and so forth.

As it turns out, characterizations of the NCC show precisely this pattern of reverberating back-and-forth communications among different brain regions (n.a. 2011).

Researchers suspect even that, when damage to the primary visual cortex presumably interrupts an instance of this kind of reverberation, patients display blindsight (Paller & Suzuki 2014); that is, the ability to correctly discriminate moving objects despite the reported inability to see them. This is precisely what one would expect if the reverberation in question were the oscillations between *X* and *N*: The objects are consciously perceived—which therefore explains how the patients discriminate them—but the patients do not know *that* they consciously perceive the objects.

By mistaking meta-consciousness for consciousness, we create two significant problems: First, we fail to distinguish between conscious processes that lack re-representation and truly unconscious processes. After all, both are equally unreportable to self and others. This misleads us to concluding that there is a mental unconscious when, in reality, there may always be something it feels like to have each and every mental process in our psyche. Second, we fail to see that our partial and tentative explanations for the alleged rise of consciousness may concern merely the rise of conscious metacognition.

This is liable to create the illusion that we are making progress toward solving the 'hard problem of consciousness' (Chalmers 2013) when, in fact, we are bypassing it altogether: mechanisms of metacognition are entirely unrelated to the problem of how the qualities of experience could arise from physical arrangements.

Consciousness may never arise—be it in babies, toddlers, children or adults—because it may always be there to begin with. For all we know, what arises is merely a metacognitive configuration of preexisting consciousness. If so, consciousness may be fundamental in nature—an inherent aspect of every mental process, not a property constituted or somehow generated by particular physical arrangements of the brain. Claims, grounded in subjective reports of experience, of progress toward reducing consciousness to brain physiology may have little—if anything—to do with consciousness proper, but with mechanisms of metacognition instead.

Chapter 27

Misreporting and Confirmation Bias in Psychedelic Research

What do images of the brain under psychedelics really tell us about its relation to the mind?

With Edward F. Kelly

(The original version of this essay was published on Scientific American *on 3 September 2018)*

A long-awaited resurgence in psychedelic research is now under way and some of its early results have been startling. Whereas most scientists expected the mind-boggling experiences of psychedelic states to correlate with increased brain activity, a landmark study from 2012 (Carhart-Harris *et al.*) found the opposite to be the case. Writing in this magazine, neuroscientist Christof Koch expressed the community's collective surprise (2012b). These unexpected findings have since been repeatedly confirmed with a variety of psychedelic agents and measures of brain activity (Muthukumaraswamy *et al.* 2013, Palhano-Fontes *et al.* 2015, Carhart-Harris *et al.* 2016, Lewis *et al.* 2017).

Under the mainstream physicalist view that brain activity is—or somehow generates—the mind, the findings certainly seem counterintuitive: How can the richness of experience go up when brain activity goes down? Understandably, therefore, researchers have subsequently endeavored to find *something* in patterns of brain activity that reliably increases in psychedelic states. Alternatives include brain activity variability (Tagliazucchi *et al.* 2014), functional coupling between different brain areas (Petri *et al.* 2014) and, most recently, a property of brain activity variously labeled as 'complexity,' 'diversity,' 'entropy' or 'randomness' (Carhart-Harris 2018)—terms viewed as approximately synonymous.

The problem is that modern brain imaging techniques do detect clear spikes in raw brain activity when sleeping subjects dream even of dull things such as staring at a statue (Costandi 2013) or clenching a hand (Hamzelou 2011). So why are only decreases in brain activity conclusively seen when subjects undergo psychedelic experiences, instead of dreams? Given how difficult it is to find *one* biological basis for consciousness (Miller 2005), how plausible is it that *two* fundamentally different mechanisms underlie conscious experience in the otherwise analogous psychedelic and dreaming states?

Perhaps because it is so hard to make sense of these results, science journalists routinely report them inaccurately, sometimes encouraged by careless statements from the researchers themselves. For instance, a 2014 study found that psychedelics increase activity *variability* in certain brain regions (Tagliazucchi *et al.*). Naturally, variability is not the same as activity, for the same reason that acceleration is not the same as speed. Yet, here is how the media reported on the study:

> Researchers ... found increased activity in regions of the brain that are known to be activated during dreaming. (IFL Science n.a.)

This echoes the way a study co-author seems to have inadvertently misrepresented the study's findings in a non-technical essay:

> [The psychedelic] increased the amplitude (or 'volume') of activity in regions of the brain that are reliably activated during dream sleep. (Carhart-Harris 2014)

Despite these statements, the technical study alluded to

(Tagliazucchi *et al.* 2014) says nothing of the kind, neither explicitly nor by implication. It only shows that activity levels *varied* more in psychedelic states.

This pattern of misreporting is consistent and sustained, as one of us elaborated upon elsewhere (Kastrup 2016c), following the spectacularly erroneous manner in which the media covered the 2016 publication of yet another study (Carhart-Harris *et al.* 2016).

Let us be clear: we are not suggesting malicious intent. Our point is that paradigmatic expectations can make it all too easy to cherry-pick, misunderstand and then misrepresent results so as to render them consistent with the reigning worldview. And because the community at large shares the same expectations, such errors easily go unnoticed.

Perhaps more worryingly, paradigmatic expectations may be playing a disproportionate role in the research itself. For example, in a 2014 study (Petri *et al.*) the correlations between activity in different brain regions were represented as graphs, with the regions as nodes and the associated correlations as links drawn between nodes. By applying successively lower levels of correlation as the minimum threshold for linking nodes, the researchers created the appearance that the brain under psychedelics displays dramatically increased global connectivity.

In all fairness, the authors themselves described these graphs as "simplified cartoons" and encouraged caution in their interpretation. However, the graphs were subsequently used with no such qualification by respected journalist Michael Pollan—in his otherwise excellent book on psychedelics (2018)—as the primary prop for a conventional physicalist interpretation of the brain imaging results. Puzzlingly, Pollan barely mentions the far more impressive and direct measurements of decreased brain activity reported in multiple other studies.

Relatively recently, researchers reanalyzed data from several experiments using their own measures of the brain activity 'diversity' mentioned earlier (Schartner *et al.* 2017). Their measures successfully discriminated between ordinary waking consciousness and conditions involving diminished awareness. More importantly, they also discriminated statistically between waking consciousness and states of expanded awareness produced by psychedelics.

However, the increases in 'diversity' observed in psychedelic states were tiny—far smaller than the reductions associated with diminished awareness—and occurred very near the top of the complexity scale, meaning that there is little room for improvement. An inherent and unresolved tension also remains between (a) 'diversity' as a measure of differentiation in neural activity and (b) the long-range integration across brain regions that is required by the associated theories (Koch *et al.* 2016). Finally, to suggest that brain activity *randomness* explains psychedelic experiences seems inconsistent with the fact that these experiences can be highly structured and meaningful—often even the most meaningful in life (Griffiths *et al.* 2006, 2008).

In short, a formidable chasm still yawns between the extraordinary richness of psychedelic experiences and the modest alterations in brain activity patterns so far observed. It remains possible that further improvements in measurement technique will at least partly bridge this chasm. If an alternative measure closely related to, but more elaborate than, 'diversity' (Casali 2013) were applied to psychedelic states, for example, a conspicuous gap in the literature would be closed.

A relatively recent overview situates the latest psychedelic research in a larger historical context (Swanson 2018). Informed observers have always interpreted psychedelic experiences as incursions into awareness of normally hidden parts of the mind. The hypothesis is that psychedelics disrupt some sort of 'reducing valve' mechanism that normally confines awareness within limits defined by the needs of everyday life. Today's physicalist neuroscientists aspire to provide an account of this process in strictly neural terms, under the assumption that everything that enters awareness must come from somewhere in the brain.

Earlier advocates of the 'reducing valve' model, however, felt compelled by evidence to adopt a broader view: consciousness, they argued, "overflows the organism" (Henri Bergson) and is ultimately grounded in some sort of transpersonal "mind at large" (Aldous Huxley).

Although determined proponents of the physicalist worldview regard such ideas as unworthy of consideration, we dare to think otherwise. For one, transformative experiences like those produced by psychedelics can occur under a wide variety of circumstances bearing little physiological similarity with psychedelic states (Timmermann *et al.* 2018). A broader—but still naturalistic—'reducing valve' model can accommodate these phenomena. It can also more naturally make sense of the overall reductions in brain activity associated not only with psychedelics—their most prominent effect—but also brain function impairment (see Chapter 25 of this book), as well as numerous related phenomena (Kelly *et al.* 2009).

Despite assumptions to the contrary, materialism is not the only scientifically grounded worldview on offer. Alternative conceptual frameworks of that sort exist, as we have elaborated upon in our books (Kastrup 2019, Kelly, Crabtree & Marshall 2015). The psychedelic brain imaging research discussed here has brought us to a major theoretical decision point as to which framework best fits with all relevant data. We hope this essay will encourage open-minded readers to take seriously the wider possibilities now coming into view.

Chapter 28

Yes, Free Will Exists

Just ask Schopenhauer

(The original version of this essay was published on Scientific American *on 5 February 2020)*

At least since the Enlightenment, in the 18th century, one of the most central questions of human existence has been whether we have free will. In the late 20th century, some thought neuroscience had settled the question (Libet 1985). However, as it has recently become clear (Gholipour 2019), such was not the case. The elusive answer is nonetheless foundational to our moral codes, criminal justice system, religions and even to the very meaning of life itself—for if every event of life is merely the predictable outcome of mechanical laws, one may question the point of it all.

But before we ask ourselves whether we have free will, we must understand what exactly we mean by it. A common and straightforward view is that, if our choices are predetermined, then we don't have free will; otherwise we do. Yet, upon more careful reflection, this view proves surprisingly inappropriate.

To see why, notice first that the prefix "pre" in "predetermined choice" is entirely redundant. Not only are all predetermined choices determined by definition, all determined choices can be regarded as predetermined as well: they always result from dispositions or necessities that precede them. Therefore, what we are really asking is simply whether our choices are determined.

In this context, a free-willed choice would be an undetermined one. But what is an undetermined choice? It can only be a random one, for anything that isn't fundamentally random reflects some underlying disposition or necessity that determines it. There is no semantic space between determinism and randomness that could accommodate choices that are neither. This is a simple but important point, for we often think—incoherently—of free-willed choices as neither determined nor random.

Our very notion of randomness is already nebulous and ambiguous to begin with. Operationally, we say that a process is random if we can't discern a pattern in it. However, a truly random process can, in principle, produce any pattern by mere chance. The probability of this happening may be small, but it isn't zero. So when we say that a process is random, we are merely acknowledging our ignorance of its potential underlying causal basis. As such, an appeal to randomness doesn't suffice to define free will.

Moreover, even if it did, when we think of free will we don't think of mere randomness. Free choices aren't erratic ones, are they? Neither are they undetermined: if I believe that I make free choices, it is because I feel that my choices are determined *by me*. A free choice is one determined by my preferences, likes, dislikes, character, etc., as opposed to someone else's or other external forces.

But if our choices are always determined anyway, what does it mean to talk of free will in the first place? If you think about it carefully, the answer is self-evident: *we have free will if our choices are determined by that which we experientially identify with.* I identify with my tastes and preferences—as consciously felt by me—in the sense that I regard them as expressions of myself. My choices are thus free insofar as they are determined by these felt tastes and preferences.

Why, then, do we think that metaphysical materialism—the notion that our choices are determined by neurophysiological activity in our own brain—contradicts free will? Because, try as we might, we don't experientially identify with neurophysiology; not even our own. As far as our conscious life is concerned, the neurophysiological activity in our brain is merely an abstraction.

All we are directly and concretely acquainted with are our fears, desires, inclinations, etc., as experienced—that is, our felt volitional states. So, we identify with these, not with networks of firing neurons inside our skull. The alleged identity between neurophysiology and felt volition is merely a conceptual—not an experiential—one.

The key issue here is one that permeates the entire metaphysics of materialism: all we ever truly have are the contents of consciousness, which philosophers call 'phenomenality.' Our entire life is a stream of felt and perceived phenomenality. That this phenomenality somehow arises from something material, outside consciousness—such as networks of firing neurons—is a theoretical inference, not a lived reality; it's a narrative we create and buy into on the basis of conceptual reasoning, not something felt. That's why, for the life of us, we can't truly identify with it.

So the question of free will boils down to one of metaphysics: Are our felt volitional states reducible to something outside and independent of consciousness? If so, there cannot be free will, for we can only identify with contents of consciousness. But if, instead, neurophysiology is merely how our felt volitional states *present themselves to observation* from an outside perspective—that is, if neurophysiology is merely the *image* of conscious willing, not its cause or source—then we do have free will; for in the latter case, our choices are determined by volitional states we intuitively regard as expressions of ourselves. I have elaborated more extensively on these admittedly nuanced and easy-to-misunderstand ideas in my earlier book, *Brief Peeks Beyond* (Kastrup 2015: 171-180), which I encourage the reader to peruse.

Crucially, the question of metaphysics can be legitimately broached in a way that inverts the usual free will equation: according to 19th-century philosopher Arthur Schopenhauer, it is the laws of nature that arise from a transpersonal will, not the will from the laws of nature. Felt volitional states are the irreducible foundation of both mind and world. Although Schopenhauer's views are often woefully misunderstood and misrepresented—most conspicuously by presumed experts—when correctly construed they offer a coherent scheme for reconciling free will with seemingly deterministic natural laws.

As elucidated in my concise book, *Decoding Schopenhauer's Metaphysics* (Kastrup 2020), for Schopenhauer the inner essence of everything is conscious volition—that is, will. Nature is dynamic because its underlying volitional states provide the impetus required for events to unfold. Like his predecessor Immanuel Kant, Schopenhauer thought of what we call the 'physical world' as merely an image, a perceptual representation of the world in the mind of an observer. But this representation isn't what the world is like *in itself*, prior to being represented.

Since the information we have about the external environment seems to be limited to perceptual representations, Kant considered the world-in-itself unknowable. Schopenhauer, however, argued that we can learn something about it not only through the sense organs, but also through *introspection*. His argument goes as follows: even in the absence of all self-perception mediated by the sense organs, we would still experience our own endogenous, felt volition. Therefore, prior to being represented we are essentially will. Our physical body is merely how our will presents itself to an external vantage point. And since both our body and the rest of the world appear in representation as matter, Schopenhauer inferred that the rest of the world, just like ourselves, is also essentially will.

In Schopenhauer's illuminating view of reality, the will is indeed free *because it is all there ultimately is*. Yet, its image is nature's seemingly deterministic laws, which reflect the instinctual inner consistency of the will. Today, over 200 years after he first published his groundbreaking ideas, Schopenhauer's work can reconcile our innate intuition of free will with modern scientific determinism.

Part VII

Broader Perspectives

Chapter 29

Metaphysics and Woo

An outsider's perspective on academic philosophy's social role

(At the time of this writing, an earlier version of this essay was scheduled for publication on the Blog of the American Philosophical Association*)*

As someone who has published a number of books on the nature of life and reality, I regularly receive emails from ordinary members of the reading public who want to share their own metaphysical theories with me. The vast majority aren't educated in philosophy: some have little education at all, whereas others have doctorates and are even widely recognized in their own fields—such as neurology, engineering, biology, etc.—but not in philosophy. I know from other authors that they, too, receive the same kind of emails.

As if this weren't enough to betray the metaphysical unrest underlying our culture today, as publisher of *Iff Books* I also regularly review philosophy manuscripts submitted by lay people. These aspiring philosophers have come to a point in their lives where their idiosyncratic metaphysical intuitions have more or less congealed. Tellingly, the resulting insights are much more satisfying to them than the mainstream views they inherit from our culture—such as 'scientific' materialism and religious dualism—which is alarming: here we have lay but reasonable people sincerely believing they can singlehandedly do better than the mainstream. Even more alarmingly, *often they actually can*.

Although materialism is the reigning metaphysics in

our Western intellectual establishment, the consensus that underlies it seems to be merely superficial; a habit rather than a conviction. At least in my own relatively large and varied circle of acquaintances, many scientists, engineers and scholars pay lip service to materialism—for a variety of pragmatic reasons—but, in their heart of hearts, harbor more than a few doubts about it. Just under the seemingly quiet surface of the *consensus gentium*, our intellectual establishment seems to be brimming with idiosyncratic metaphysical perspectives.

In his impressive and important book, *A Secular Age* (2007), Charles Taylor calls this proliferation of metaphysical views the "nova effect." It began as a reaction to the erosion of religious faith in the nineteenth century and is now—with the advent of the Internet and social media—reaching bizarre proportions. In Taylor's words, the effect consists in the spawning of

> an ever-widening variety of moral/spiritual options, across the span of the thinkable and perhaps even beyond. ... The fractured culture of the nova, which was originally that of elites only, [now] becomes generalized to whole societies. (2007: 299)

And thus, philosophers no longer have a monopoly on creative metaphysical thinking. Or rather, *everybody is now a philosopher*, even if not properly trained as one.

Indeed, there is an important sense in which doing philosophy is intrinsic to being human. With our unique capacity for reflection, we not only *can*, but perhaps are *destined to*, ask the big questions: Who are we? What's going on? What's the point of it all? From this perspective, doing philosophy is much more than just the day job of philosophy graduates. With the possible exception of those who must struggle constantly to merely survive, the rest of us inevitably face these questions at some point in life. Hence, we all philosophize, haphazardly as the case may be.

And here is where the active participation of *trained, professional* philosophers in our cultural dialogue has never been more important in the whole history of the West. When it comes to metaphysics, our culture's confusion is unparalleled, which is both cause and self-fulfilling consequence of the nova effect. The undeniable success of technology has even encouraged some scientists—typically the self-appointed, militant spokespeople of 'scientific' materialism—to conflate what *works* with what *is*, and thus conclude that behavior-predicting models can replace philosophical reflection.

Only lucid, clear, persuasive philosophy can break this vicious cycle. Not only are the self-appointed spokespeople of 'scientific' materialism unwittingly creating a metaphysical mess of historical proportions, the average educated person is trying to fill the gap and do some philosophy on their own. Without professional help to identify and demarcate the field of plausible play, what chance do they have? One by one, they fall prey to common logical fallacies that have been known for at least two and a half thousand years.

Indeed, there is an important sense in which philosophy—like science and engineering—is an acquired skill, not only an art that rests on innate intuition and inspiration. We don't trust people who never studied medicine to perform an operation, or people who never trained to be pilots to fly an airplane; why should we expect people with no training in philosophy to do good philosophy, no matter how innately gifted they are? To think straight and clearly isn't something we are born knowing. The seductive pitfalls of logical fallacies are many and often exceedingly subtle, hard to discern. Those who have studied philosophy—at least informally—have seen many examples of these fallacies and are thus better equipped to spot and avoid them. They are also much less prone to naively trying to reinvent the wheel by pursuing avenues of thinking that have already been extensively explored and debated before.

Therefore, professional philosophy has an indispensable role to play in society. And insofar as exploring philosophical questions is intrinsic to our shared humanity, I even dare to claim that philosophy is one of the most important professions; if not *the* most important. Whereas more economically prestigious fields such as medicine, law, engineering, etc., are merely utilitarian—that is, means to an end—philosophy directly addresses our reasons to live. Doctors may help us stay alive, lawyers may help us stay out of jail, engineers may make our lives more comfortable, convenient and fun; but only philosophers address what we live *for*.

Yet, if we look around, do we honestly see philosophers playing such a decisive role in our society? If not, why not? The social need for professional, lucid philosophizing is blatant and overwhelming, so what has gone wrong?

For starters, philosophers can learn a thing or two from the spokespeople of scientific materialism: they have managed not only to pass science for metaphysics, but also to make science look exciting and fun. The media isn't blind to it, and thus science gets a lot of airtime, a lot of opportunities to influence our culture and the lives of our fellow citizens. Philosophy, on the other hand, seems largely stuck in an ivory tower of impenetrable jargon, ghostly abstractions and unending hair-splitting. How have we come to find ourselves in this situation?

It's not too difficult to see how, if we are honest with ourselves. Imagine the popular appeal of a TV show featuring two analytic philosophers discussing and trying to pin down the exact meaning of, say, the concept of causality. Would it be comparable to the appeal of a show featuring, say, the UK's baby-faced rock star physicist Brian Cox speaking unintelligibly about quantum entanglement, drawing 'Whoa!s' from a befuddled but captivated audience? Honestly, we may have something to learn from that kind of thing.

Academic philosophy's failure to arouse comprehensive popular interest doesn't change the fact that lucid philosophy is what we need most in this bewildering "nova" era of ours. The philosophy ship carrying us through life has lost its trained officers and is now at the mercy of winds and currents. The alarming growth in cases of anxiety, depression, ennui and despair we are witnessing is—I am convinced—largely a symptom of the unsustainable lack of firm metaphysical foundations in our culture. After all, it is not easy to find oneself in the strange position of being alive, constantly fighting against entropy, knowing that one day one is guaranteed to lose the fight. If we philosophers don't help people make sense of, and peace with, this condition, the pharma industry will continue to fill the void.

However, to properly play that role we must attune our work to the needs of society, as opposed to playing navel-gazing conceptual games that get papers published yet achieve little else. Philosophy is not meant for philosophers alone, otherwise it should be a club, not a publicly funded human activity. If we are to be relevant, philosophers must also actively communicate with the public, like scientists do. The responsibility for making the significance of our work clear to everybody can only be our own. Communication is an integral part of the job, not just something you throw over the wall for somebody else to worry about. Corporate managers know this well, because they've learned it the hard way.

There are, of course, valiant attempts to reach a wider public with some popular philosophy publications. And these are crucially important, something to build on. But are these efforts comparable to the popular science media industry? In 2020, *Scientific American* completed 175 years in continuous publication, the longest run of *any* magazine, not only scientific ones. It faced competition from *New Scientist*, *Pop Science*, *Cosmos*, *Discover*, *American Scientist*, *National Geographic*, etc. *Psychology*

Today is read by a broad and incredibly diverse audience. There are multiple subscription-based TV channels dedicated to science. PBS's science channels on YouTube gather millions of subscribers. Clearly, scientists have something figured out that we philosophers, in our insular world, ignore at our own peril.

The need for sober philosophy in society is overwhelming; perhaps even bigger than the need for science. But we cannot hope to emulate the public relations success of science unless professional philosophy drives to *conclusions*—tentative as they may be—as opposed to increasingly convoluted, open-ended debates. Science, too, revises its views every so often, sometimes in ways that contradict everything it viewed as certain before: just think of how relativity and quantum mechanics turned previous theories on their head. *But at any one point in history science does offer tentative conclusions,* the best-guesses of the time. And this is what allows science to inform and influence society so extensively as it does. Philosophy, too, must adopt such attitude: accept that conclusions are always tentative and may eventually change, *but do drive to conclusions*.

Moreover, the tendency in philosophy to create broad comparative taxonomies of metaphysical options—instead of making *choices*—is of dubious value, for if the social role of philosophy is to provide lucid guidance to navigating the metaphysical undercurrents of our culture, the internal relativism of the philosophy community defeats its own *raison d'être*: we can't help the average person figure out in which direction to point the rudder if we insist that all directions may make sense from some vantage point.

Granted, philosophy often can't appeal directly to experiment—as science almost always can—to settle disputes. However, we still have internal logical consistency, conceptual parsimony and empirical adequacy as criteria to evaluate our metaphysical options. Albeit somewhat subjective, these are criteria about which there is a fair level of consensus. What is missing is our collective willingness to consistently apply them and own the consequences of doing so.

Far from me to pretend that I have ready-made solutions or patronize the academic philosophy community, in which I am an outsider. But perhaps it is precisely this outside perspective that allows me to contribute a different angle. In this spirit, I believe the key is in realizing that in both science and business—two worlds in which I earned my living for many years—the opportunity to hold on to ultimately untenable views is finite; sooner or later *reality* comes calling and settles the questions for the better or worse. But in philosophy, there seems to be endless opportunity to espouse one's own pet view and go largely unpunished for it.

I submit that the professional philosophy community must find or devise some Darwinian process to play, in metaphysics, a role analogous to experiments in science and market performance in business. Without something like it the work of philosophers shall remain forever divided, open-ended and irrelevant from a social perspective. For philosophers cannot truly influence the cultural dialogue without collectively taking some (tentative) position—or at least discarding others—as opposed to relativizing everything under tortuously abstract taxonomies.

With the march of A.I., automation and other megatrends that render mechanical, repetitive human labor increasingly redundant, the philosophy profession—along with art and spirituality the only human activity that directly tackles the meaning of life—has everything to become the cornerstone of our society. But for that to happen, we, philosophers, must first get our collective act together.

Chapter 30

The Conceivability Trap

Analytic philosophy's Achilles' Heel

(The original version of this essay was published on the Blog of the American Philosophical Association *on 14 May 2020)*

The possibility of objective knowledge is a contentious topic that continues to be debated. However, no reasonable practitioner will deny that, in practice, both science and analytic philosophy *aim* for objectivity: methods of data collection, analysis and synthesis that lead to conclusions at least largely independent of subjective perspective. Moreover, in practice most scientists and analytic philosophers implicitly believe that their conclusions are, by and large, objective; for this is often how they present their findings in the technical literature.

Since our empirical experiences are always perspectival—after all, each of us operates through a unique point of view or window into the world—the achievement of objective knowledge is contingent upon a procedure meant to distil objectivity out of perspectival subjectivity. In science, Karl Popper offered the following: "the objectivity of scientific statements lies in the fact that they can be *inter-subjectively tested*" (2005, emphasis added). In analytic philosophy, however, issues often cannot be settled by experimental testing, so a different form of procedural objectivity is required. In this context, Bertrand Russell held that there are *a priori* principles of logical reasoning—*not* contingent on the idiosyncratic perspectives inherent to empirical experience—which, if properly applied, render objective conclusions possible (1995, 2009).

Even philosophers of mind, whose object of study is that

most subjective of all things, aim for objectivity. Expressions now common in the community—such as 'what it is like to be (something or someone)' to define the presence of experience—as well as words such as 'phenomenal' and 'access' to qualify consciousness, reflect an effort to objectify what is essentially subjective.

But can the ideal of full objectivity ever be realized? For as long as analytic philosophers are fallible human beings, instead of computers, it surely can't. Their conclusions, too, are inevitably a function of the variety and metacognitive depth of their personal experiences. It is more productive to acknowledge this fact and respond accordingly, than to pretend otherwise.

For instance, the notion of *conceivability*—which is often appealed to in modern ontology and philosophy of mind to establish or refute metaphysical possibility—relies on the particular set of subjective experiences a philosopher has had in his or her life. Therefore, it is naïve—perhaps even pretentious—to assume that one's personal inability to conceive of something entailed or implied by an argument positively refutes the argument. For not only in continental, but also analytic philosophy, one's conclusions reveal perhaps as much about oneself as they do about one's object of inquiry.

Indeed, even language itself—an indispensable tool not only for communicating, but also formulating our thoughts—is based on shared *experience*. Words only have meaning to us insofar as their denotations and connotations are experiences we've had ourselves. For instance, because you and I have experienced a car, the word 'car' has meaning for both of us, and so we can use it in a conversation. Similarly, because the word 'color' denotes an experience I've had, I can use the concept of color in my own meditations about the nature of mind.

As a matter of fact, the concept of a color *palette* occupies centerstage today in philosophy of mind. Analytic philosophers who adhere to constitutive panpsychism use the concept to *conceive*—by analogy—of how a limited set of fundamental phenomenal states could be combined—like pigments in a palette—to constitute our ordinary experiences. The *conceivability* of this very notion rests on our *shared experience* of colors and how they can be combined to form other colors.

Now imagine a person born blind becoming an analytic philosopher. The person doesn't share with sight-capable philosophers the experience of having mixed watercolors in kindergarten. As a matter of fact, the person doesn't even know what a color is. The very notion of a palette of fundamental experiences that could be combined to form meta-experiences wouldn't be conceivable *to the person*. And yet, the rest of us knows it is perfectly conceivable. Conceivability is thus not an objective notion, but an inherently subjective one.

The trap of conceivability in philosophy of mind can come in much more subtle, nuanced and elusive—yet no less perilous—forms than phenomenal palettes. To demonstrate it, I shall use an example very close to my heart, for it happens to be the reason why many of my colleagues hesitate about my own ideas regarding the nature of mind and reality.

Whereas constitutive panpsychism faces the so-called 'combination problem' (i.e. how can micro-level phenomenal subjects combine to form macro-level subjects such as you and me?) my formulation of idealism—'analytic idealism'—faces the so-called 'decomposition problem': How can one *universal* subject ground our *personal, seemingly separate* subjectivities? How can the one ground the many?

I address this problem with the largely empirical notion of dissociation: in psychiatry, we know that the mind of a person suffering from dissociative identity disorder can apparently fragment itself, leading to the formation of disjoint personalities or 'alters.' The reality of alter formation has been demonstrated with modern neuroimaging research over the last decade (see Chapter 14 of this book). Therefore, even if we don't know exactly *how* it happens, we do know *that* it happens.

Each alter is a seemingly separate subject within the host mind and can be conscious at the same time as other alters, even during dreams (Barrett 1994: 170-171). These extraordinary clinical observations give us more than just a hint for how to solve the decomposition problem on empirical grounds, as opposed to theoretical abstractions. Yet, it can be very challenging for many of us to *conceive* of how one subject can seemingly break up and form disjoint but *co-conscious* sub-subjects.

Let me phrase this challenge in a clearer manner: as philosopher Sam Coleman put it, a subject "can be thought of as a point of view annexed to a private qualitative field" (2014). Dissociation thus implies (a) the existence of *more than one point of view within the same mind,* (b) each point of view being annexed to a *different qualitative field,* (c) the qualitative fields being experienced *simultaneously* yet (d) *separately.* How are we to conceive of this?

Steps (a) to (c) can be easily conceived of by analogy to a simple phenomenological experiment I originally proposed in a book (Kastrup 2014). Place your right indicator finger, pointing up, at an arm's-length distance from your face, in the middle of your visual field (go on, have a go at it). Looking at it with both of your eyes, you see a single image: your indicator finger pointing up.

Now close one of your eyes while keeping the other open. Then alternate between the two eyes, so you always have only one eye open at a time, looking at your finger. You will notice that the image from each eye is only slightly different from that of the other.

But if you slowly bring your finger closer to your face, while continuing to alternate between your eyes, you will notice that the difference grows. When the finger finally touches your nose, you get very different images when looking at it with your left or your right eye; so different, in fact, that the part of your finger visible to each eye is not at all visible to the other. This gives us (a) *two points of view* within the same mind (*your* mind), (b) each annexed to a *different qualitative field*.

If you now open both of your eyes with the finger still touching your nose, you will see an overlay of both images: an attempt by your mind to *merge* them together. In other words, you are now (c) *simultaneously conscious* of two different qualitative fields, each experienced from a different point of view. However, the respective experiences aren't (d) *separate*, because your brain tries to merge the corresponding fields in a single image.

Full dissociation is what would happen if, instead of merging the two qualitative fields, there were one 'you' experiencing the world through your right eye, and another, *separate* but concurrently conscious 'you' experiencing the same world through your left eye. This would give us (d) and complete the exercise in conceivability.

Alas, conceiving of (d) explicitly and coherently is contingent upon one's idiosyncratic life experiences: for some it is straightforward, whereas for others it is so difficult they consider the whole notion of co-conscious alters fundamentally incoherent. The psychiatric evidence that these alters, however they form, somehow *do* form isn't sufficient for the latter group, so indispensable conceivability seems to be.

And now we've come full circle: to be able to *conceive* of dissociative processes leading to seemingly disjoint but co-conscious subjects, one must have had the *experience* of dissociation oneself; at least in a mild, non-pathological form. Insofar as this experience isn't widely shared across the analytic philosophy community, consensus about dissociation being the solution to the decomposition problem shall remain elusive. So much for objectivity.

It is impossible for me or anyone else to directly talk *about* the felt qualities of dissociation, because they are not common enough to be part of our culture-bound, shared dictionary of experiences. But I can try to talk *around* them, in the hope that you pick them up with your peripheral vision.

More than once in my life, after my 35th year, I became suddenly cognizant of certain feelings and emotions—even thoughts—that I had been experiencing for most of my life but hadn't been explicitly aware of. When this kind of realization occurs, you tell yourself, "I've always known this," or "I've always felt this way, but wasn't aware of it." You realize *retroactively* that there was a disjoint part of you, separate from the executive ego, which *experienced* feelings, emotions and thoughts inaccessible to the ego due to lack of associative bridges.

Let me be clear: at the moment of the realization, you *know* that the dissociated feelings, emotions and thoughts in question were being experienced *continuously* and therefore *concurrently* with the executive ego, by a *subject* dissociated from the ego. After this subject is reintegrated with the ego, you realize that *both were you*, all along. And it is by virtue of this reintegration—which includes the memories of each of the subjects—that you know, retroactively and with absolute clarity, that you had been two: you can now experience the memories of *both* subjects as your own memories, including the memory that each of the subjects couldn't access the other's phenomenal field for as long as the dissociation lasted.

Anyone who has had this experience can conceive of dissociation leading to subject decomposition. As a matter of fact, those who have had this experience *know* that subject decomposition isn't merely theoretical; *it actually happens.*

Although presumably not a typical experience for analytic philosophers—perhaps due to the character traits, dominant cognitive functions and innate dispositions that lead them to select their profession in the first place—what I described above is often recognizable among practitioners of mindfulness, depth psychology, nondual self-inquiry and a few other related fields in which much attention is paid to the metacognitive representation of one's own phenomenal states. The metaphysical possibility of subject decomposition through dissociation would be more easily accepted there.

There is, thus, an important sense in which analytic philosophy is epistemically handicapped by a degree of structural disregard for phenomenology and self-reflective introspection (at least beyond the abstract and objectifying conceptualizations that constitute its day-to-day practice). This self-imposed and unnecessary limitation results from the somewhat naïve belief that analytic philosophy can be done in a fully objective manner: to preserve the mere illusion of full objectivity, we dismiss important *subjective* sources of knowledge. Indeed, even science may have something to gain from not holding as tightly to the phantasm of full objectivity: "attempts to rid science of perspectives might not only be futile because scientific knowledge is necessarily perspectival, they can also be epistemically costly because they prevent scientists from having the epistemic benefits certain standpoints afford," said Julian Reiss and Jan Sprenger (2017).

In philosophy of mind, lack of attention to phenomenology and self-reflective introspection carries a particularly high cost. For if we turn the study of mind into a purely abstract conceptual process—in which concepts are deliberately separated from our lived experiences, like cards on a table—we will lose touch with the very target of our inquiry. By trying to *objectively* inquire into mind—that most *subjective* of all things—we fall into a fundamental, epistemically confining contradiction. After all, isn't it self-evident that, to properly study the mind, one must try to know one's own mind?

Not only continental but analytic philosophy too, whether it recognizes it or not, is fundamentally dependent on *personal, direct experience* to make progress, for conceivability is a function of the life we've lived, the richness and depth of the experiences we've had. As the love of wisdom, philosophy—like wisdom—grows out of lived experience, not mere abstraction.

Chapter 31

The Meaning and Destiny of Western Culture

Thoughts on Peter Kingsley's oeuvre and its relation to my work

(The original version of this essay was published on Science & Nonduality *on 12 March 2020)*

One of the more salient intellectual events of 2019 for me, personally, was my discovery of the work of Peter Kingsley. Unlike most of the books I read—which I tend to regard rather soberly—Kingsley's work left me irate, inspired, bemused and delighted, all at the same time. I am anything but indifferent to it, which is probably the greatest compliment I could pay to any author. The implication of Kingsley's argument is that non-dualism and idealism aren't purely Eastern insights, but the metaphysical and spiritual root of the West as well. This is what I set out to discuss in this essay.

31.1 A culture's source and telos

Kingsley's central premise is that all cultures have a sacred source and purpose, including our own Western civilization: "everything, absolutely everything, anyone can name that makes our so-called civilization unique has a sacred source—a sacred purpose" (Kingsley 2018: 228). The seed of every culture, including our own, is planted through visionary experience. It is prophets who learn, and then inform us of, what our purpose is: "Western civilization, just like any other, came into being out of prophecy; from revelation" (*Ibid.*: 231).

In our case, we can trace our roots back to visionary Greek

philosopher-poets living in southern Italy about two and a half thousand years ago, particularly Parmenides. In Parmenides' poem *On Nature* we can find the origins of our Western culture.

31.2 Misunderstanding Parmenides

However, Kingsley argues that we have been *mis*interpreting and *mis*representing Parmenides' ideas since Plato. Parmenides is considered by mainstream scholars to be the founder of logic and rationality, of our particular way of *discriminating* truth from untruth, fact from fiction, through *reasoning*. According to this mainstream view, the Promethean powers of Western science, as embodied in technology, are the culmination of a way of thinking, feeling and behaving that can be traced back to Parmenides' manner of argumentation in his famous poem.

But Kingsley argues very persuasively (2003: 1-306) that what Parmenides was trying to say was nothing of the kind. According to him, logic for Parmenides wasn't a formal system based on fixed axioms and theorems, meant to help us *discern* true from false ideas about reality; it wasn't grounded in some metaphysically primary realm of absolutes akin to Platonic Forms; it didn't derive its validity from some external reference. In summary, Kingsley argues that, for Parmenides, logic wasn't what we now call *reason*, but something much broader, deeper, unconstrained by fixed rules and formalisms.

31.3 True logic as incantation

As a matter of fact, according to Kingsley Parmenides' logic was a kind of *incantation*. The context is the notion that we live in a world of illusions, caught up in our own internal narratives and made-up categories about what is going on, completely oblivious to the true world that surrounds us and from which we derive our very being—i.e. *reality*. This illusion is persuasive, has tremendous power and momentum. So to help one see through it and ultimately overcome it, an even *more persuasive* rhetorical device is required, a kind of spell or incantation woven with words, meant to disrupt our ordinary mental processes. This incantation is the *true* logic Parmenides gifted us, a kind of spell meant to trick our internal storytelling, make it catch itself in contradiction and thereby release its grip, so we can escape the illusion.

This is a critical point, so allow me to belabor it a bit. If I were to use Parmenides' true logic on you, I would weave whatever argument line I felt would be compelling *to you*, irrespective of whether the argument is strictly rational or not, strictly consistent with a given set of fixed axioms or not. The ultimate goal of true logic is eminently pragmatic: it is to get you out of the bind in which you continuously put yourself. True logic, thus, is a semantic trick meant to break the spell of illusion, like cracking a crystal by gently tapping on it in just the right spot.

31.4 Parmenides' metaphysics

Kingsley explains that, for Parmenides, there were only two ways to approach reality: either we decide that everything that is felt, thought, perceived, imagined or otherwise experienced exists *as such*—regardless of any correspondence with ostensibly objective facts—or we must ultimately dismiss everything as non-existing. The latter option goes nowhere, for obvious reasons, which leaves only one viable path. The bind we find ourselves in is due to our hopeless attempt to find some compromise or middle ground between those two canonical options: we try to *discriminate* which mental states correspond to actual existents—i.e. to some external reference—and which don't. *This,* according to Kingsley's interpretation of Parmenides, *is the core of the illusion.* And true logic is a rhetorical tool meant to show that all such discriminations—if pursued consistently to their final implications—are ultimately self-defeating.

The implicit metaphysics being adopted here is, of course, idealism: "for Greeks, the world of the gods [i.e. reality] had one very particular feature. This is that simply to think something is to make it exist: is to make it real" (Kingsley 2003: 71-72). Therefore, "whatever we are aware of is, whatever we perceive or notice is, whatever we think of is" (*Ibid.*: 77). Everything that has mental existence *exists as such*—i.e. as a *mental* existent—*and there is no other way in which it can exist*: "There is nothing that exists except what can be thought or perceived" (*Ibid.*: 78). Therefore, the use of reason to discriminate between what exists from what doesn't exist is ultimately *unreasonable*: "To choose good thoughts is to reject the bad ones—and to reject something is to entertain it, is to make it exist" (*Ibid.*: 80). The act of deciding that something does not, or cannot, exist immediately backfires and *makes it exist*, by the mere fact that the act forces us to think it into existence to begin with. Reason, as we normally apply it, is thus ultimately incoherent, even though it has many practical applications within the context of the illusion.

It is the idealism he attributes to Parmenides that renders Kingsley's interpretation internally consistent: once a world ostensibly outside consciousness—not necessarily your or my seemingly personal consciousness alone, but outside *transpersonal* consciousness—is done away with, all criteria of truth and existence become *internal* ones, and thus logic boils down to *persuasion*: what exists or is true is whatever consciousness has been persuaded to *make* exist or true. There is nothing outside consciousness to make it otherwise.

Kingsley explains: "facts are of absolutely no significance in themselves: it's just as easy to get lost in facts as it is to get lost in fictions. ... All our facts, like all our reasoning, are just a façade" (*Ibid.*: 21-22), they hide something more essential behind them. And this 'something' is *reality*: pure stillness, a realm in which nothing ever moves or changes, in which everything is intrinsically connected to everything else in an indivisible whole, and where no time but the eternal present exists. That's why true logic is "a magical lure drawing us into oneness" (*Ibid.*: 144)—i.e.

back to reality.

31.5 Reason is not true logic

Kingsley explains that, because we have historically misinterpreted and misrepresented Parmenides' intended meaning, we've ended up conjuring up reason out of what was meant to be true logic. *But reason is a tool precisely for discriminating between mental states that ostensibly correspond to facts outside consciousness from those that don't.* Under the metaphysical view that to think is to make exist, such discrimination is incoherent.

Therefore, by misunderstanding true logic, we've also departed from what was meant to be Western culture's foundational metaphysics. We've invented external references outside consciousness—i.e. outside *reality*—such as matter, energy, space and time. And then we've forced true logic "to operate, distorted and disfigured, in the world it had been designed to undermine" (Kingsley 2003: 144)! The result is *reason*, the rational discrimination of fact from fiction in an ostensibly autonomous material world.

For Kingsley, it is reason that keeps us stuck in the middle ground between the two canonical paths—namely, between deciding either that everything that is conceived in consciousness exists *as such*, or that nothing exists. This, according to him, is the seminal mistake that has put our entire culture on the wrong footing. Logic is no longer regarded as a magical incantation meant to persuade us out of illusion, but has turned into a tool for perpetuating the illusion: "All our attempts to discriminate between reality and deception or between truth and illusion are exactly what keeps on tricking us"(Ibid.: 211).

31.6 The telos of Western culture

But what was it that we were originally supposed to do? What goal are we supposed to pursue? What is the "burning purpose at the heart of our Western world" (Kingsley 2018: 205)?

Kingsley is not terribly explicit about it, but he does drop enough hints. For instance, he says that the modern attitude towards the divine can be summarized in the words,

> "Let's make sure the divine takes good care of us. But as for finding what, in reality, the divine might possibly need: let it look after itself." From here onwards one can sit back and watch how the idea of looking after the gods starts, almost by magic, vanishing from the Western world. ... And now it never for a moment occurs to us that the divine might be suffering, aching from our neglect; that the sacred desperately longs for our attention far more than we in some occasional, unconscious spasm might feel a brief burst of embarrassed longing for it. (*Ibid.*: 29-30)

The suggestion is that the meaning and purpose of our lives is to help fulfill some divine need, which can only be fulfilled in, or by means of, the state of consciousness we call life. This is reinforced by the fact that Kingsley overtly associates himself with the thought of Swiss psychiatrist Carl Jung, particularly Jung's book *Answer to Job*. And in that book, we find Jung saying:

> what does man possess that God does not have? Because of his littleness, puniness, and defencelessness against the Almighty, he possesses ... a somewhat keener consciousness based on self-reflection: he must, in order to survive, always be mindful of his impotence. God has no need of this circumspection, for nowhere does he come up against an insuperable obstacle that would force him to hesitate and hence make him reflect on himself. (2002b: 14-15)

It seems to me that *all* cultures have the purpose to serve the divine by means of the state of consciousness we call life, the latter not being available to the divine itself. But each culture is meant to fulfill this sacred task in its own particular way, according to its own particular dispositions and strengths. In the case of Western culture, our strength is our sharply developed *meta-cognition*, or *self-reflection*: our introspective ability to turn our own thoughts, emotions, perceptions and fantasies into objects of thought, recursively. Western culture is thus meant to serve the divine by contributing to it the meta-cognitive insight of *self-realization*: through us and our Western science—"a gift offered by the gods with a sacred purpose" (Kingsley 2018: 229)—the divine recognizes itself.

31.7 The failure of the West

However, Kingsley ultimately concludes that we, in the West, have failed in our divine task. We've failed not only because we've misunderstood Parmenides—and thus bungled our metaphysics and become unable to properly use the sacred tools we were given, namely, true logic and science—but for other, more insidious reasons as well.

Indeed, to serve the divine requires "a deeply religious attitude, the sense that it's all for the sake of something far greater than ourselves" (Kingsley 2018: 122). But to nurture and sustain such religious attitude, people must "step out of their personal dramas" (*Ibid.*). Yet we, in the West, indulge in personal dramas, having conflated individual freedom and expression with egocentrism, even subtle forms of narcissism. We've forgotten that, "as humans we are archetypes" (*Ibid.*: 143), instances of a universal template of being, so that "Whatever we think of as personal is in fact profoundly inhuman, while it's only in the utter objectivity of the impersonal that we find our humanity" (*Ibid.*).

Worse yet, Kingsley maintains that there is no fixing the problem, no rescuing Western culture, no finding our path again: "this world of ours is already dead. It existed for a while, did the best it could, but is nothing more than a lifeless remnant of what it was meant to be. ... And this is the moment for marking, and honouring, the passing of our culture ... to keep on indulging in optimism is a shameless dereliction of our duty" (Ibid.: 442).

Well, I am not an optimist ... But I don't agree.

31.8 The *de facto* Western culture and the value of error

The first thing to notice is that, although Kingsley has convinced me that we did misinterpret Parmenides, and that the correct interpretation is that offered by Kingsley, the fact of the matter is that what we call 'Western culture' embodies and is based on the values, premises and modes of cognition set by Plato, Aristotle and the rest of the post-Socratic philosophers and scientists. According to Kingsley himself, Parmenides was misinterpreted already within a single generation, so there has never been a 'correct' Western culture, so to speak. Factually, even if it is based on a seminal misunderstanding, being Western effectively means what Plato and his successors defined it to be; it has never really meant anything else.

Moreover, I don't think that the seminal errors of the West were a waste of time either. Wisdom sometimes comes only with error, as any wounded healer will know. Sometimes a misstep is more useful and important than the correct way forward, because of the experiences and insights it creates the space for. Getting to the right answer only after having exhaustively tried*, and failed with,* seductive but wrong ones arguably leads to a deeper, fuller insight than getting things right first-time round. For in the former case, one is more intimately acquainted with why and how those seductive answers are actually illusory, and therefore has an equally fuller comprehension of the right answer.

More specifically, by having embraced objective facts and reasoning fully, unreservedly, we are making sure that every stone is turned, particularly the most seductive ones; we are laying the ground for a deeper future insight than what those shooting straight for the end can achieve. The destiny of Western culture may entail experimenting with extremely seductive but deluded answers first, exhausting the alternatives, and only then setting itself straight. Of course, the price we pay for this is unfathomable. Generation upon generation have endured grief, despair, unspeakable suffering of every kind for having followed the siren song of illusion. This is the West's sacrifice. The only question is whether we will eventually get it right or not.

31.9 Prison break

But just *how* can we eventually get out of this bind and unveil *reality?* Kingsley talks often about μῆτις (*mêtis*), a kind of cunning wisdom that can be used to trick, enchant or persuade. The illusion we live in is a product of μῆτις, and only *more persuasive* μῆτις, such as true logic, can get us out of it.

Now ask yourself: What would be *truly persuasive* for the Western mind? What kind of story could short-circuit our internal narratives, expose their inner contradictions and force us to review our unexamined assumptions? The answer seems absolutely crystal clear to me: *reasoning consistently pursued to its ultimate implications.*

The Western mind acknowledges only reasoning as a valid story. It will dismiss anything else without even looking at it. So if one wants to use true logic to trick the West out of illusion, this true logic *must* come disguised as reason; it must entail *embracing the illusion fully*, objective facts and all, and judiciously applying reason within it. That's the μῆτις required here; there's just no other way.

To free the West from illusion, we must first break into the prison wherein the West finds itself, and then break out again carrying the rest of the culture with us. We must fight the duel with the weapons chosen by the opposition, for those are the only weapons the opposition recognizes as real. Kingsley himself is well aware of this approach: "there are methods that reality can use to work its own way into our illusion and start to draw us out" (Kingsley 2003: 255). Ditto. What a fantastic movement of μῆτις it would be to use pure, strict, sharp reasoning to undermine reason itself.

31.10 Transcending reason through reasoning

And, as it turns out, if pursued to its ultimate, final implications, reason *does* undermine and relativize itself. Through reasoning we can demonstrate, in multiple redundant ways, that reason isn't absolute; that, although applicable and useful in many situations, it is relative, a convenient invention, not a fact of reality etched into stone. As a matter of fact, I've written a whole book about it, *Meaning in Absurdity* (Kastrup 2012).

The relativity of reason isn't a new insight. The West has been refining it for a long time, at least since Agrippa's famous trilemma, also known as the 'Münchhausen trilemma.' Modern scholars such as Graham Priest have further developed the associated insights (2006). In the 20th century, Kurt Gödel demonstrated that no axiomatic system—such as e.g. arithmetic—can be both complete and sound: they either fail to express every truth about themselves, or they contradict themselves (1931). And since physics—our very description of the universe—is based on axiomatic systems, a fundamental limitation seems to be established for the ability of reason to comprehend reality in both a sound and complete manner. Finally, the insights of Quantum Mechanics in the early 20th century led to a long and deep academic debate about the ground of logic (cf. e.g. Putnam 1968, Dummett 1976): Is it empirical? Is it invented? Where does it come from anyway? All these developments illustrate how far reason can be brought in undermining itself from within, under strictly rational conditions.

I discuss all this and much more in *Meaning in Absurdity*, my first foray into *true* logic. I invite interested readers (and Kingsley) to peruse it. Here is a passage from the book:

> It is ironic that science, through the diligent and consequent pursuit of a materialist, strongly-objective view of nature, would lead to the very evidence that renders such view untenable. As we will later see, it is a recurring theme in different branches of science and philosophy that the pursuit of a rational system of thought ultimately leads to its own defeat. There is something perennial about the idea that any literal view of nature, when pursued to its ultimate ramifications, destroys itself from within. It is as though every literal model carried within itself the seeds of its own falsification; as if nature resisted attempts to be limited or otherwise boxed in. Whatever we say it is, it indicates it is not; whatever we say it is not, it shows it might just be. These are built-in mechanisms of growth and renewal in nature that we ignore at our own peril. Nature is as fluid and elusive as a thought. Indeed, *it is a thought*: an unfathomable, compound thought we live in and contribute to. The world is a shared 'dream.' In it, as in a regular dream, the dreamer is himself the subject and the object; the observer and the observed. (Kastrup 2012: 44)

The Western path towards transcending reason is through the strictest, most consequent pursuit of reason possible.

31.11 Metaphysics first

However, the realization that our reason ultimately undermines itself from within could be destabilizing for Western culture, unless and until we grok the fact that all existence unfolds in consciousness. I say this because only under idealism—as opposed to materialism—can we reasonably accommodate the understanding that reason, like anything else, is but a mental construct. The embrace of idealism is the first and indispensable step in the West's return to reality. Only then can we accept the following statements made by Kingsley (in which I've added clarifying comments between brackets):

> "The moment you accept every single thought as equally true [for, if all is in consciousness, to be true *is* to be thought] and also see the truth of this, then all thinking fades into unimportance" (Kingsley 2003: 74);
>
> "There is nothing that exists except what can be thought or perceived [for there is nothing beyond mentation]" (*Ibid.*: 78);
>
> "Absolutely everything, including the fabric of reality itself, is trickery and illusion [i.e. consciousness deceiving itself by believing the products of its own imagination, without which there would be precisely *nothing* except the mere potential for experience]" (Ibid.: 91);
>
> "There is deception at the heart of reality, and the other way around [for, in consciousness, to exist *is* to be imagined and then believed, like a dream we believe to be true while we are in it]"(Ibid.: 211);
>
> "Everybody is a myth. You are a myth [for your very sense of individual identity is a story you tell yourself, in consciousness, and then believe it]" (*Ibid.*: 158);
>
> "All that exists is now [for in consciousness, as I discussed in Chapter 22 of this book, the past are memories experienced now, and the future are expectations experienced now]" (*Ibid.*: 164).

Embracing idealism within the constraints of the game of reason is a necessary first step in our path forward; it is the step that creates the space required for all other steps.

31.12 Concepts versus experience

One could say that understanding and embracing idealism is merely an abstract conceptual game, which isn't transformative. Conceptual conclusions don't sink into the body, but instead circle around in our head as loops of thought; they don't change much the way we feel and behave. Only *direct experience* is transformative, for it percolates throughout our entire being. To know what reality is, one must experience it directly, not only grasp it intellectually. Otherwise one stays stuck in mere descriptions—like a would-be traveler who only knows places by the information in brochures—and never becomes *acquainted* with what it is all about. In Kingsley' words: "Until we have the direct experience of reality we are ... totally helpless. We can't understand a thing" (Kingsley 2003: 255).

And I concur. Only direct experience is transformative. However, given the mindset of Western culture, one must first give oneself intellectual *permission* to have the experience, to be rationally open to it, in order to have any chance of experiencing it. One must be conceptually primed to accept the experience if and when it comes, for otherwise our rational defense mechanisms will instinctively and promptly shut ourselves off from it. It is critical that we first bring down our defenses through μῆτις suitable for the prickly Western mind—i.e. reasoning—because the intellect is the bouncer of the heart. In the West, what the intellect dismisses as impossible, or nonsense, or woo, or flakey, bounces off our head and never sinks in.

This is why embracing idealism as a metaphysics is a crucial first step in the West. We must first give ourselves intellectual permission to experience what is currently considered impossible or nonsensical, for only then will we truly *recognize* and *accept* the experience when it comes.

31.13 Beyond idealism

As a matter of fact, it is plausible that, even without direct experience, we could grasp some of the more counterintuitive characteristics of reality that Kingsley describes. For instance: "Cunning and trickery ... are woven into the fabric of the universe. Everything around us is an elaborate trick" (Kingsley

2003: 219); or "the origin of the universe is now" (*Ibid.*: 169); or "everything is one, whole, motionless" (*Ibid.*: 255); etc. If one has intellectually bought into idealism, these seemingly contradictory and even outright absurd statements can be made understandable through suitable argumentation; suitable *incantations* that gently hold the intellect by the hand and take it beyond the boundaries of its domain; a kind of μῆτις much more subtle and delicate than that required to argue for idealism itself; in summary: *true logic.*

Arguments based on true logic need to skirt and transcend the edge of rationality, exceed the envelop of strict, explicit, unambiguous reasoning. They are really a kind of conceptual spell meant to take one beyond conceptual thinking. And it is extraordinarily difficult to compose them correctly, for the slightest fault brings down the whole building.

For instance, it is true that reality is constructed out of belief; pure belief, nothing else; if there is no belief, there is nothing. But if one is to make this statement and leave it at that, one is bound to be misinterpreted and dismissed. For we will fall and die if we jump off a building, *even if we believe we can fly*; the world doesn't seem at all acquiescent to our beliefs. The point here, however, isn't that reality is constituted by *personal, egoic* beliefs; the foundational beliefs in question aren't accessible through introspection; they underly not only a person, not only a species, not only all living beings, but *everything*. They aren't *our* beliefs, but the beliefs that bring us into being in the first place.

Another example: As Kingsley says, trickery is woven into the fabric of the universe. This is entirely true, but if he or I were to just leave it at that, the intellect of our readers would trounce the statement: obviously the physical world is just *natural*, it is doing merely what it is compelled to do by natural *laws*; it isn't the product of trickery by some god up in the sky. The actual point, however, is different: since reality unfolds in consciousness, and consciousness is also its own witness, the only way for things to feel real is if consciousness tricks itself into believing that its own imagination is an external phenomenon. *Consciousness's prime directive is to trick itself, for if it doesn't, nothing is left but a void.* That's the point.

And yet, there is much more to these points. This 'more' isn't at all easy to describe in words in a manner that wouldn't sound totally foolish and self-contradictory. So an important part of the story is always missing from reasoning. In some cases, even the *whole* story is missing.

For instance, a few times in my life readers have confronted me on the implications of analytic idealism, one of which is that consciousness—*your* consciousness—never dies. They say: "People are dying all the time, how can you deny that?" To which I sometimes reply, in an effort of μῆτις: "Have you ever noticed that only *other* people die, never you?" When the reader stops to think about this answer, it inevitably fails to convey any meaningful insight. The reader might justifiably say to me: "Of course, you idiot, for I can only notice that other people die while I am alive myself!" But if the reader's conceptual armor is momentarily caught off guard and the answer sneaks past the reasoning mind, it can unlock a deep but completely ineffable insight; an insight that is impossible to capture *in* words, but which is nonetheless recognized as *unequivocally true*. I can't *explain* it here; I can only gently tap the crystal that constitutes your conceptual armor and hope that it cracks, so you can see past it.

That's the challenge facing authors who want to go beyond rationality, to unveil a little more of reality than what can be corralled into explicit and unambiguous concepts. It really requires a kind of incantation or spell: words that don't capture the message *in* themselves, in an explicit and reasonable manner, but somehow tap on something in the reader's mind that allow a sudden flood of intuitions to rush in. This is what the enchantment of *true logic* can achieve if done correctly.

I have tried to do it in print. This is what my book *More Than Allegory* is all about (Kastrup 2016a). Despite its subtitle ('*On religious myth, truth and belief*') it is really a book about *reality*; about aspects of reality that can't be captured by analytic philosophy. In the book, I use *true* logic to try and convey ideas that transcend reasoning. And yet, I attempt to render these ideas in a manner friendly to rationality—i.e. I try to help the reader go beyond the intellect in a way that isn't threatening to the intellect; that doesn't scare off our conceptual reasoning but, instead, makes an ally of it. Indeed, I try to make ultimately unreasonable ideas as reasonable-sounding as possible. This is the book's μῆτις.

To give you a sense of how I went about this challenge, here is a passage from the book wherein I touch on the subject matter of the following statements by Kingsley's: "Cunning and trickery ... are woven into the fabric of the universe" (Kingsley 2003: 219), "the origin of the universe is now" (*Ibid.*: 169), and "everything is one, whole, motionless" (*Ibid.*: 255):

> Despite its intangibility, all of existence must fit within the present moment, for the present is all there ever is. Even the past and the future, as myths experienced in the present, exist within it. Thus, out of the quasi-nothingness of the now somehow comes everything. ... The present moment is the cosmic egg described in many religious myths ... It is a singularity that births all existence into form. It seeds our mind with fleeting consensus images that we then blow up into the voluminous bulk of projected past and future. These projections are like a cognitive 'big bang' unfolding in our mind. They stretch out the intangibility of the singularity into the substantiality of events in time. But unlike the theoretical Big Bang of current physics, the cognitive 'big bang' isn't an isolated occurrence in a far distant past. It

happens now; now; now. It only ever happens now. ... Existence only appears substantial because of our intellectual inferences, assumptions, confabulations and expectations. What is actually in front of our eyes now is incredibly elusive. The volume of our experiences—the bulk of life itself—is generated by our own internal myth-making. We conjure up substance and continuity out of sheer intangibility. We transmute quasi-emptiness into the solidity of existence through a trick of cognitive deception where we play both magician and audience. In reality, *nothing ever really happens*, for the scope of the present isn't broad enough for any event to unfold objectively. That we think of life as a series of substantial happenings hanging from a historical timeline is a fantastic cognitive hallucination. Roger Ebert's last words, illuminated by the clarity that only fast-approaching death can bring, seem to describe it most appropriately: "This is all an elaborate hoax." And who do you think is the hoaxer? (Kastrup 2016a: 102-104)

31.14 Looking forward

The value of Kingsley's work for me has been the precise opposite of what he overtly tried to achieve: instead of convincing me that the West is dead and must be mourned, I now have renewed faith that it is not only alive, but viable. Perhaps this was, all along, Kingsley's covert intention with the book. For nothing motivates people like me more than facing a contrarian attitude; nothing mobilizes more energy for action than being told that our efforts are hopeless.

We, authors, are slaves to our daemons—instigating inner spirits with an autonomy and agenda of their own—which is symbolized by the circular chain in my coat of arms. My own daemon is particularly ruthless, so I couldn't just stop my work even if Kingsley or anyone else had convinced me, intellectually, that there is no point to it. I just can't stop. But importantly, the fact that my daemon is more energetic than ever after I read Kingsley suggests to me that *there is still hope*. Perhaps Kingsley's own daemon tricked him (daemons are masters of this sort of trickery): by announcing the death of Western culture, Kingsley may have inadvertently revitalized it; prompted me and many others to redouble our efforts to prove that this isn't the end, that there is still much to be done. Maybe that was the plan of Kingsley's daemon all along ...

I emerge from my in-depth engagement with Kingsley's thoughts with more clarity regarding the role my various books play in a broader historical and cultural context. Some of them—*Rationalist Spirituality* (2011a), *Why Materialism Is Baloney* (2014), *Brief Peeks Beyond* (2015) and *The Idea of the World* (2019)—comply fully with the premises and constraints of rational thought, strict reasoning, aiming to convince you that idealism is the most reasonable interpretation of reality. Others—*Dreamed up Reality* (2011b), *Meaning in Absurdity* (2012) and *More Than Allegory* (2016a)—are instances of true logic: they seek to use reasoning to transcend reasoning, to help you glimpse certain mental landscapes or insights that cannot be captured in explicit and unambiguous words.

The West is alive, it only looks lost. I know it because my daemon knows it. I am myself a quintessential embodiment of Western rationalism: have the highest academic degree in both the sciences and the humanities, from two of Europe's top universities; have been raised and educated immersed in Western thinking; have worked in some of the most recognizable Western scientific institutions; have earned my living in the cut-throat world of Western high-tech business; the life streams of my ancestors—my own dead—from Northern, Southern and Western Europe meet in the river of my veins and life. *And still, despite all this, I recognize where Kingsley is coming from* (or at least I *think* I do); I am not lost (hopefully). So if I take myself as a representative example—which is the only thing I can do, since

I don't have access to other people's inner lives—the West is, just under the surface, still very vital. We do have a future and a destiny to fulfill.

Onwards we go.

Bibliography

Ackroyd, E. (1993). *A Dictionary of Dream Symbols*. London, UK: Cassell Illustrated.

Adelson, EH (1995). Checker shadow illusion. [Online]. Available from: http://persci.mit.edu/gallery/checkershadow [Accessed 12 April 2017].

Adelson, R. (2005). Hues and views: A cross-cultural study reveals how language shapes color perception. *American Psychological Association*, 6 (2): 26. [Online]. Available from: https://www.apa.org/monitor/feb05/hues [Accessed 22 April 2020].

Albert, H. (1985). *Treatise on Critical Reason*. Princeton, NJ: Princeton University Press.

American Psychiatric Association (2013). *Diagnostic and Statistical Manual of Mental Disorders* (5th ed.). Washington, DC: American Psychiatric Publishing.

Ananthaswamy, A. (2011). Quantum magic trick shows reality is what you make it. *New Scientist*, June 22. [Online]. Available from: https://www.newscientist.com/article/dn20600-quantum-magic-trick-shows-reality-is-what-you-make-it/ [Accessed 14 June 2016].

Anderson, PW (1972). More is different. *Science*, 4 August, 177 (4047): 393-396.

Ansmann, M. *et al.* (2009). Violation of Bell's inequality in Josephson phase qubits. *Nature*, 461: 504-506. DOI: 10.1038/nature08363.

Aspect, A., Grangier, P. and Roger, G. (1981). Experimental tests of realistic local theories via Bell's theorem. *Physical Review Letters*, 47 (7): 460-463.

Aspect, A., Dalibard, J. and Roger, G. (1982). Experimental test of Bell's inequalities using time-varying analyzers. *Physical Review Letters*, 49 (25): 1804-1807.

Aspect, A., Grangier, P. and Roger, G. (1982). Experimental realization of Einstein-Podolsky-Rosen-Bohm gedankenexperiment: A new violation of Bell's inequalities. *Physical Review Letters*, 49 (2): 91-94.

Augusto, LM (2010). Unconscious knowledge: A survey. *Advances in Cognitive Psychology*, 6: 116-141.

Barbour, J. (1999). *The End of Time: The Next Revolution in Physics*. Oxford, UK: Oxford University Press.

Barfield, O. (2011). *Saving the Appearances: A Study in Idolatry*. Oxford, UK: Barfield Press.

Barrett, D. (1994). Dreams in dissociative disorders. *Dreaming*, 4 (3): 165-175.

Bell, J. (1964). On the Einstein Podolsky Rosen Paradox. *Physics*, 1 (3): 195-200.

BIG Bell Test Collaboration, The (2018). Challenging local realism with human choices. *Nature*, 557: 212-216. DOI: https://doi.org/10.1038/s41586-018-0085-3.

Black, DW and Grant, JE (2014). *The Essential Companion to the Diagnostic and Statistical Manual of Mental Disorders* (5th ed.). Washington, DC: American Psychiatric Publishing.

Blanke, O. *et al.* (2002). Stimulating illusory own-body perceptions: The part of the brain that can induce out-of-body experiences has been located. *Nature*, 419: 269-270.

Block, N. (1995). On a confusion about a function of consciousness. *Behavioral and Brain Sciences*, 18: 227-287.

Bohm, D. (1952a). A suggested interpretation of the quantum theory in terms of "hidden" variables. I. *Physical Review*, 85: 166-179.

Bohm, D. (1952b). A suggested interpretation of the quantum theory in terms of "hidden" variables. II. *Physical Review*, 85: 180-193.

Bohm, D. (1980). *Wholeness and the Implicate Order*. London, UK: Routledge.

Boly, M. *et al.* (2011). Preserved feedforward but impaired top-down processes in the vegetative state. *Science*, 332 (6031): 858-862.

Boswell, J. (1820). *The Life of Samuel Johnson, LL. D.* (Volume 1). London, UK: J. Davis, Military Chronicle and Military Classics Office.

Braude, S. (1995). *First Person Plural: Multiple Personality and the Philosophy of Mind*. New York, NY: Routledge.

Brenner, ED *et al.* (2006). Plant neurobiology: An integrated view of plant signaling. *Trends in Plant Science*, 11 (8): 413-419.

Buks, E. *et al.* (1998). Dephasing in electron interference by a 'which-path' detector. *Nature*, 391: 871-874.

Buonomano, D. (2018). *Your Brain Is a Time Machine: The Neuroscience and Physics of Time*. New York, NY: WW Norton & Company.

Burke, BL, Martens, A. and Faucher, EH (2010). Two decades of terror management theory: A meta-analysis of mortality salience research. *Personality and Social Psychology Review*, 14 (2): 155-195.

Carhart-Harris, RL *et al.* (2012). Neural correlates of the psychedelic state as determined by fMRI studies with psilocybin. *Proceedings of the National Academy of Sciences of the United States of America*, 109 (6): 2138-2143.

Carhart-Harris, RL (2014). Magic mushrooms expand your mind and amplify your brain's dreaming areas – here's how. *The Conversation*, 3 July. [Online]. Available from: https://theconversation.com/magic-mushrooms-expand-your-mind-and-amplify-your-brains-dreaming-areas-heres-how-28754 [Accessed 21 April 2020].

Carhart-Harris, RL *et al.* (2016). Neural correlates of the LSD experience revealed by multimodal neuroimaging. *Proceedings of the National Academy of Sciences of the United States of America* (PNAS Early Edition), doi: 10.1073/pnas.1518377113.

Carhart-Harris, RL (2018). The entropic brain – revisited. *Neuropharmacology*, 142: 167-178. DOI: https://doi.

org/10.1016/j.neuropharm.2018.03.010.
Carruthers, P. and Schechter, E. (2006). Can panpsychism bridge the explanatory gap? *Journal of Consciousness Studies*, 13 (10-11): 32-39.
Cartwright, J. (2007). Quantum physics says goodbye to reality. *IOP Physics World*, April 20. [Online]. Available from: http://physicsworld.com/cws/article/news/2007/apr/20/quantum-physics-says-goodbye-to-reality [Accessed 14 June 2016].
Casali, AG *et al.* (2013). A theoretically based index of consciousness independent of sensory processing and behavior. *Science Translational Medicine*, 5 (198). DOI: 10.1126/scitranslmed.3006294.
Catani, D. (2013). *Evil: A History in Modern French Literature and Thought*. London, UK: Bloomsbury.
Chalmers, D. (1996). *The Conscious Mind: In Search of a Fundamental Theory*. Oxford, UK: Oxford University Press.
Chalmers, D. (2003). Consciousness and its place in nature. In: Stich, S. & Warfield, T. (eds.). *The Blackwell Guide to Philosophy of Mind*. Malden, MA: Blackwell.
Chalmers, D. (2006). Strong and weak emergence. In: Davies, P. and Clayton, P. (eds.). *The Re-Emergence of Emergence*. Oxford, UK: Oxford University Press.
Chalmers, D. (2016a). The combination problem for panpsychism. In: Brüntrup, G. & Jaskolla, L. (eds.). *Panpsychism*. Oxford, UK: Oxford University Press.
Chalmers, D. (2016b). Panpsychism and panprotopsychism. In: Brüntrup, G. & Jaskolla, L. (eds.). *Panpsychism*. Oxford, UK: Oxford University Press.
Chalmers, D. (2018). Idealism and the mind-body problem. In: Seager, W. (ed.). *The Routledge Handbook of Panpsychism*. London, UK: Routledge.
Cheetham, T. (2012). *All the World an Icon: Henry Corbin and the Angelic Function of Beings*. Berkeley, CA: North Atlantic Books.
Chen, S. (2019). Sean Carroll thinks we all exist on multiple

worlds: In his book *Something Deeply Hidden*, the physicist explores the idea of Many Worlds, which holds that the universe continually splits into new branches. *WIRED*, 10 September. [Online]. Available from: https://www.wired.com/story/sean-carroll-thinks-we-all-exist-on-multiple-worlds/ [Accessed 26 April 2020].

Cleeremans, A. (2011). The radical plasticity thesis: How the brain learns to be conscious. *Frontiers in Psychology*, 2, article 86.

Coleman, S. (2014). The real combination problem: Panpsychism, micro-subjects, and emergence. *Erkenntnis*, 79 (1): 19-44.

Costandi, M. (2013). Brain scans decode dream content: Researchers have decoded the content of people's dreams using brain scanning technology. *The Guardian*, 5 April. [Online]. Available from: https://www.theguardian.com/science/neurophilosophy/2013/apr/05/brain-scans-decode-dream-content [Accessed 21 April 2020].

Coyne, J. (2020). Muddled philosopher: Consciousness could not have evolved. *Why Evolution Is True* (Jerry Coyne's personal blog), 7 February. [Online]. Available from: https://whyevolutionistrue.wordpress.com/2020/02/07/philosopher-consciousness-could-not-have-evolved/ [Accessed 24 April 2020].

Crabtree, A. (2009). Automatism and Secondary Centers of Consciousness. In: Kelly, EF *et al. Irreducible Mind: Toward a Psychology for the 21st Century*. Lanham, MD: Rowman & Littlefield.

Cristofori, I. *et al.* (2016). Neural correlates of mystical experience. *Neuropsychologia*, 80: 212-220.

Crowther, K. (2014). *Appearing Out of Nowhere: The Emergence of Spacetime in Quantum Gravity*. Doctoral Dissertation, The University of Sidney.

Damasio, A. (2011). Neural basis of emotions. *Scholarpedia*, 6 (3): 1804.

Dehaene, S. and Changeux, J-P (2011). Experimental and theoretical approaches to conscious processing. *Neuron,* 70 (2): 200-227.

Dennett, DC (1991). *Consciousness Explained.* London, UK: Penguin Books.

Dennett, DC (1995). *Darwin's Dangerous Idea: Evolution and the meanings of life.* New York, NY: Simon & Schuster.

Dennett, DC (1996). The scope of natural selection. *Boston Review,* October/November: 34-38.

Dick, PK (2001). *Valis.* London, UK: Gollancz.

Dijksterhuis, A. and Nordgren, LF (2006). A theory of unconscious thought. *Perspectives on Psychological Science,* 1 (2): 95-109.

DiSalvo, D. (2012). When you inject spirit mediums' brains with radioactive chemicals, strange things happen. *Forbes,* November 18. [Online]. Available from: http://www.forbes.com/sites/daviddisalvo/2012/11/18/when-you-inject-spirit-mediums-brains-with-radioactive-chemicals-some-really-strange-things-happen/ [Accessed 25 February 2017].

Dummett, M. (1976). Is Logic Empirical? In: Lewis, HD (ed.). *Contemporary British Philosophy, 4th series.* London, UK: Allen and Unwin, pp. 45-68.

Eagleman, DM (2009). Brain time. In: Brockman, M. (ed.). *What's Next? Dispatches on the Future of Science.* New York, NY: Vintage.

Eagleman, DM (2011). *Incognito: The Secret Lives of the Brain.* New York, NY: Canongate.

Eddington, AS (1928). *The Nature of the Physical World.* New York: The Macmillan Company.

Einstein, A. (1921). Geometry and experience: Address to the Prussian Academy of Sciences, 27 January. London, UK: Methuen & Co. Ltd. [Online]. Available from: http://mathshistory.st-andrews.ac.uk/Extras/Einstein_geometry.html [Accessed 26 April 2020].

Einstein, A., Podolsky, B. and Rosen, N. (1935). Can quantum-mechanical description of physical reality be considered complete? *Physical Review,* 47: 777-780.

Eliade, M. (2009). *Rites and Symbols of Initiation: The Mysteries of Birth and Rebirth.* New York, NY: Spring Publications.

Emerging Technology from the arXiv (2019). A quantum experiment suggests there's no such thing as objective reality. *MIT Technology Review,* 12 March. [Online]. Available from: https://www.technologyreview.com/s/613092/a-quantum-experiment-suggests-theres-no-such-thing-as-objective-reality/ [Accessed 22 April 2020].

Engel, GS *et al.* (2007). Evidence for wavelike energy transfer through quantum coherence in photosynthetic systems. *Nature,* 446: 782-786. DOI: 10.1038/nature05678.

Feser, E. (2016). Omnibus of fallacies. *First Things,* February. [Online]. Available from: https://www.firstthings.com/article/2016/02/omnibus-of-fallacies [Accessed 24 April 2020].

Fields, C. *et al.* (2017). Eigenforms, interfaces and holographic encoding: Toward an evolutionary account of objects and spacetime. *Constructivist Foundations,* 12 (3): 265-291.

Floridi, L. (2008). Modern trends in the philosophy of information. In: Adriaans, P. and Benthem, J. van (eds.). *Handbook of the Philosophy of Science, Volume 8: Philosophy of Information.* Amsterdam, The Netherlands: Elsevier, pp. 113-131.

Fonagy, P. *et al.* (eds.) (2012). *The Significance of Dreams.* London, UK: Karnac Books.

Ford, BJ (2010). The secret power of the single cell. *New Scientist,* 206 (2757): 26-27.

Fraassen, BC van (1980). *The Scientific Image.* Oxford, UK: Oxford University Press.

Fraassen, BC van (1990). *Laws and Symmetry.* Oxford, UK: Oxford University Press.

Frankish, K. (2019). The consciousness illusion: Phenomenal consciousness is a fiction written by our brains to help us track the impact that the world makes on us. *Aeon.*

[Online]. Available from: https://aeon.co/essays/what-if-your-consciousness-is-an-illusion-created-by-your-brain [Accessed 25 April 2020].

Frankish, K. (2020). The demystification of consciousness: Illusionists don't deny that consciousness exists, but propose that we rethink what it is. *IAI News*, 20 March. [Online]. Available from: https://iai.tv/articles/the-demystification-of-consciousness-auid-1381 [Accessed 26 April 2020].

Frankl, VE (1991). *The Will to Meaning* (Expanded ed.). New York, NY: Meridian.

Franklin, SP (1997). *Artificial Minds*. Cambridge, MA: MIT Press.

Franz, M-L von (1964). The process of individuation. In: Jung, CG (ed.). *Man and His Symbols*. New York, NY: Anchor Press.

Franz, M-L von (1974). *Number and Time: Reflections Leading Toward a Unification of Depth Psychology and Physics*. Evanston, IL: Northwestern University Press.

Franz, M-L von and Boa, F. (1994). *The Way of the Dream*. Boston, MA: Shambhala Publications.

Fredkin, E. (2003). An introduction to digital philosophy. *International Journal of Theoretical Physics*, 42 (2): 189-247.

Fredkin, E. (n.a.). *On the Soul (draft)*. [Online]. Available from: http://www.digitalphilosophy.org//wp-content/uploads/2015/07/on_the_soul.pdf [Accessed 2 July 2016].

Friston, K. (2013). Life as we know it. *Journal of the Royal Society Interface*, 10 (86): 20130475.

Friston, K., Sengupta, B. and Auletta, G. (2014). Cognitive dynamics: From attractors to active inference. *Proceedings of the IEEE*, 102 (4): 427-445.

Gaal, S. van *et al.* (2011). Dissociable brain mechanisms underlying the conscious and unconscious control of behavior. *Journal of Cognitive Neuroscience*, 23 (1): 91-105.

Gabrielsen, P. (2013). When does your baby become conscious? *Science*, April 18. [Online]. Available from: http://www.sciencemag.org/news/2013/04/when-does-your-baby-become-conscious [Accessed 10 September 2017].

Gholipour, B. (2019). A famous argument against free will has been debunked: For decades, a landmark brain study fed speculation about whether we control our own actions. It seems to have made a classic mistake. *The Atlantic*, 10 September. [Online]. Available from: https://www.theatlantic.com/health/archive/2019/09/free-will-bereitschaftspotential/597736/ [Accessed on 21 April 2020].

Gillespie, A. (2007). The social basis of self-reflection. In: Valsiner, J. and Rosa, A. (eds.). *The Cambridge Handbook of Sociocultural Psychology*. New York, NY: Cambridge University Press, pp. 678-691.

Glasersfeld, E. von (1987). An introduction to radical constructivism. In: Watzlawick, P. (ed.). *The Invented Reality*. New York, NY: WW Norton & Company.

Gödel, K. (1931). Über formal unentscheidbare Sätze der Principia Mathematica und verwandter Systeme, I. *Monatshefte für Mathematik und Physik*, 38 (1): 173-198. DOI: 10.1007/BF01700692.

Godfrey-Smith, P. (2014). *Philosophy of Biology*. Princeton, NJ: Princeton University Press.

Goff, P. (2006). Experiences don't sum. *Journal of Consciousness Studies*, 13 (10-11): 53-61.

Goff, P. (2009). Why Panpsychism doesn't help us explain consciousness. *Dialectica*, 63 (3): 289-311.

Goldhill, O. (2018). The idea that everything from spoons to stones is conscious is gaining academic credibility. *QUARTZ*, 27 January. [Online]. Available from: https://qz.com/1184574/the-idea-that-everything-from-spoons-to-stones-are-conscious-is-gaining-academic-credibility/ [Accessed 26 April 2020].

Graziano, M. (2016). Consciousness is not mysterious. *The Atlantic*, January 12. [Online]. Available from: http://www.theatlantic.com/science/archive/2016/01/consciousness-color-brain/423522/ [Accessed 26 February 2017].

Graziano, M. (2019). True nature of consciousness: Solving the biggest mystery of your mind: Far from being a mystical 'ghost in the machine,' consciousness evolved as a practical mental tool and we could engineer it in a robot using these simple guidelines. *New Scientist*, 18 September.

Graziano, M. (2020). Why you don't know your own mind: Rethinking consciousness and eliminativism. *IAI News*, 20 January. [Online]. Available from: https://iai.tv/articles/why-you-dont-know-your-own-mind-auid-1297 [Accessed 24 April 2020].

Greene, B. (2003). *The Elegant Universe: Superstrings, Hidden Dimensions, and the Quest for the Ultimate Theory*. New York, NY: WW Norton & Company.

Griffin, D. (1998). *Unsnarling the World-Knot*. Eugene, OR: Wipf & Stock.

Griffiths, RR *et al.* (2006). Psilocybin can occasion mystical-type experiences having substantial and sustained personal meaning and spiritual significance. *Psychopharmacology*, 187 (3): 268-283.

Griffiths, RR *et al.* (2008). Mystical-type experiences occasioned by psilocybin mediate the attribution of personal meaning and spiritual significance 14 months later. *Journal of Psychopharmacology*, 22 (6): 621-632. DOI: https://doi.org/10.1177/0269881108094300.

Gröblacher, S. *et al.* (2007). An experimental test of non-local realism. *Nature*, 446: 871-875.

Gu, M. *et al.* (2008). More really is different. *arXiv:0809.0151 [cond-mat.other]*. [Online]. Available from: https://arxiv.org/abs/0809.0151 [Accessed 26 April 2020].

Haikonen, PO (2003). *The Cognitive Approach to Conscious Machines*. Exeter, UK: Imprint Academic.

Hameroff, S. (2006). Consciousness, neurobiology and quantum mechanics: The case for a connection. In: Tuszynski, J. (ed.).

The Emerging Physics of Consciousness. Berlin, Germany: Springer-Verlag.

Hamzelou, J. (2011). Dreams read by brain scanner for the first time. *New Scientist*, 27 October. [Online]. Available from: https://www.newscientist.com/article/dn20934-dreams-read-by-brain-scanner-for-the-first-time/ [Accessed 21 April 2020].

Harris, S. (2012a). *Free Will*. New York, NY: Free Press.

Harris, S. (2012b). *Science on the Brink of Death*. Sam Harris's Blog, November 11. [Online]. Available from: https://samharris.org/science-on-the-brink-of-death/ [Accessed 21 April 2020].

Harris, S. (2016). You are more than your brain! *Big Think*, September 4. [Online]. Available from: https://www.facebook.com/BigThinkdotcom/videos/10153879575418527/ [Accessed 1 November 2016].

Hart, DB (2017). The illusionist. *The New Atlantis*, Summer/Fall issue: 109-121.

Hart, J. (2013). Toward an integrative theory of psychological defense. *Perspectives on Psychological Science*, 9 (1): 19-39.

Hartnett, K. (2017). Secret link uncovered between pure math and physics. *Quanta Magazine*, 1 December. [Online]. Available from: https://www.quantamagazine.org/secret-link-uncovered-between-pure-math-and-physics-20171201/ [Accessed 26 April 2020].

Hassin, R., Uleman, J. and Bargh, J. (eds.) (2005). *The New Unconscious*. New York, NY: Oxford University Press.

Hassin, RR (2013). Yes it can: On the functional abilities of the human unconscious. *Perspectives on Psychological Science*, 8 (2): 195-207.

Heflick, NA *et al.* (2015). Death awareness and body-self dualism: A why and how of afterlife belief. *European Journal of Social Psychology*, 45 (2): 267-275.

Heine, SJ, Proulx, T. and Vohs, KD (2006). The meaning maintenance model: On the coherence of social motivations. *Personality and Social Psychology Review*, 10 (2): 88-110.

Henry, RC (2005). The mental Universe. *Nature*, 436: 29.

Hensen, B. *et al.* (2015). Loophole-free Bell inequality violation using electron spins separated by 1.3 kilometres. *Nature*, 526: 682-686.

Hilgard, E. (1977). *Divided Consciousness*. New York, NY: John Wiley & Sons.

Hoffman, DD (2009). The interface theory of perception: Natural selection drives true perception to swift extinction. In: Dickinson, S. *et al.* (eds.). *Object Categorization: Computer and Human Vision Perspectives*. Cambridge, UK: Cambridge University Press.

Hoffman, DD and Singh, M. (2012). Computational evolutionary perception. *Perception*, 41 (9): 1073-1091.

Horgan, T. and Potrč, M. (2000). Blobjectivism and indirect correspondence. *Facta Philosophica*, 2 (2): 249-270.

Hossenfelder, S. (2010). At the frontier of knowledge. *arXiv:1001.3538 [physics.pop-ph]*. [Online]. Available from: https://arxiv.org/abs/1001.3538v1 [Accessed 26 April 2020].

Hossenfelder, S. (2020). Why the foundations of physics have not progressed for 40 years: Physicists face stagnation if they continue to treat the philosophy of science as a joke. *IAI News*, 8 January. [Online]. Available from: https://iai.tv/articles/why-physics-has-made-no-progress-in-50-years-auid-1292 [Accessed 26 April 2020].

Husserl, E. (1970). *The Crisis of European Sciences and Transcendental Phenomenology: An Introduction to Phenomenological Philosophy*. Evanston, IL: Northwestern University Press.

Huyghe, P. and Parreno, P. (2003). *No Ghost Just a Shell*. Cologne, Germany: Verlag der Buchhandlung Walther König.

IFL Science (n.a.). Your brain on magic mushrooms. [Online]. Available from: https://www.iflscience.com/brain/your-brain-magic-mushrooms/ [Accessed on 21 April 2020].

Immordino-Yang, MH *et al.* (2009). Neural correlates of admiration and compassion. *Proceedings of the National*

Academy of Sciences of the United States of America, 106 (19): 8021-8026.

Jaskolla, LJ and Buck, AJ (2012). Does panexperiential holism solve the combination problem? *Journal of Consciousness Studies*, 19 (9-10): 190-199.

Ji, Z. *et al.* (2020). MIP* = RE. *arXiv:2001.04383 [quant-ph]*. [Online]. Available from: https://arxiv.org/abs/2001.04383 [Accessed 26 April 2020].

Joos, E. (2006). The emergence of classicality from quantum theory. In: Clayton, P. and Davies, P. (eds). *The Re-Emergence of Emergence: The emergentist hypothesis from science to religion*. Oxford, UK: Oxford University Press.

Jung, CG (1985). *Synchronicity: An acausal connecting principle*. London, UK: Routledge.

Jung, CG (1991). *The Archetypes and the Collective Unconscious* (2nd ed.). London, UK: Routledge.

Jung, CG (1995). *Memories, Dreams, Reflections*. London, UK: Fontana Press.

Jung, CG (2001). *Modern Man in Search of a Soul*. New York, NY: Routledge.

Jung, CG (2002a). *Dreams*. London, UK: Routledge.

Jung, CG (2002b). *Answer to Job*. London, UK: Routledge.

Jung, CG (2014). *Analytical Psychology: Its Theory and Practice* (2nd ed.). London, UK: Routledge.

Jung, CG and Pauli, W. (2001). *Atom and Archetype: The Pauli/Jung Letters 1932-1958*. London, UK: Routledge.

Kafatos, M. and Nadeau, R. (1990). *The Conscious Universe: Part and Whole in Modern Physical Theory*. New York, NY: Springer-Verlag.

Kanai, R. (2020). Does consciousness have a function? *OUPblog*, 20 January. [Online]. Available from: https://blog.oup.com/2020/01/does-consciousness-have-a-function/ [Accessed 24 April 2020].

Karunamuni, ND (2015). The five-aggregate model of the mind.

SAGE Open, 5 (2), doi:10.1177/2158244015583860.

Kastrup, B. (2011a). *Rationalist Spirituality*. Winchester, UK: Iff Books.

Kastrup, B. (2011b). *Dreamed up Reality*. Winchester, UK: Iff Books.

Kastrup, B. (2012). *Meaning in Absurdity*. Winchester, UK: Iff Books.

Kastrup, B. (2014). *Why Materialism Is Baloney*. Winchester, UK: Iff Books.

Kastrup, B. (2015). *Brief Peeks Beyond*. Winchester, UK: Iff Books.

Kastrup, B. (2016a). *More Than Allegory*. Winchester, UK: Iff Books.

Kastrup, B. (2016b). What neuroimaging of the psychedelic state tells us about the mind-body problem. *Journal of Cognition and Neuroethics*, 4 (2): 1-9.

Kastrup, B. (2016c). The LSD study: you're being subtly deceived (again). *Metaphysical Speculations*, 12 April. [Online]. Available from: https://www.bernardokastrup.com/2016/04/the-lsd-study-youre-being-subtly.html [Accessed 21 April 2020].

Kastrup, B. (2016d). The Physicalist Worldview as Neurotic Ego-Defense Mechanism. *SAGE Open*, 6 (4). DOI: 10.1177/2158244016674515.

Kastrup, B. (2017a). Self-transcendence correlates with brain function impairment. *Journal of Cognition and Neuroethics*, 4 (3): 33-42.

Kastrup, B. (2017b). An ontological solution to the mind-body problem. *Philosophies*, 2 (2), doi:10.3390/philosophies2020010.

Kastrup, B. (2017c). On the plausibility of idealism: Refuting criticisms. *Disputatio*, 9 (44): 13-34.

Kastrup, B. (2017d). There is an 'unconscious,' but it may well be conscious. *Europe's Journal of Psychology*, 13 (3): 559-572.

Kastrup, B. (2017e). Making sense of the mental universe. *Philosophy and Cosmology*, 19: 33-49.

Kastrup, B. (2018a). The universe in consciousness. *Journal of*

Consciousness Studies, 25 (5-6): 125-155.

Kastrup, B. (2018b). Conflating abstraction with empirical observation: The false mind-matter dichotomy. *Constructivist Foundations*, 13 (3): 341-361.

Kastrup, B. (2019). *The Idea of the World*. Winchester, UK: Iff Books.

Kastrup, B. (2020). *Decoding Schopenhauer's Metaphysics*. Winchester, UK: Iff Books.

Kastrup, B. (2021). *Decoding Jung's Metaphysics*. Winchester, UK: Iff Books.

Kay, AC *et al.* (2010). Religious belief as compensatory control. *Personality and Social Psychology Review*, 14 (1): 37-48.

Kelly, EF *et al.* (2009). *Irreducible Mind: Toward a Psychology for the 21st Century*. Lanham, MD: Rowman & Littlefield.

Kelly, EF, Crabtree, A., Marshall, P. (2015). *Beyond Physicalism: Toward Reconciliation of Science and Spirituality*. New York, NY: Rowman & Littlefield.

Kihlstrom, JF (1997). Consciousness and me-ness. In: Cohen, J. and Schooler, JW (eds.). *Scientific Approaches to Consciousness*. Mahwah, NJ: Lawrence Erlbaum Associates, pp. 451-468.

Kihlstrom, J. and Cork, R. (2007). Anesthesia. In: Velmans, M. & Schneider, S. (eds.). *The Blackwell Companion to Consciousness*. Oxford, UK: Blackwell.

Kim, Y-H *et al.* (2000). A delayed 'choice' quantum eraser. *Physical Review Letters*, 84: 1-5.

Kingsley, P. (2004). *Reality*. Point Reyes Station, CA: The Golden Sufi Center.

Kingsley, P. (2018). *Catafalque*. London, UK: Catafalque Press.

Klein, SB (2015). The feeling of personal ownership of one's mental states: A conceptual argument and empirical evidence for an essential, but underappreciated, mechanism of mind. *Psychology of Consciousness: Theory, Research, and Practice*, 2 (4): 355-376.

Klimov, PV *et al.* (2015). Quantum entanglement at ambient

conditions in a macroscopic solid-state spin ensemble. *Science Advances*, 1 (10), e1501015.

Kluge, I. (n.a.). Drive on through: Sam Harris's Free Will. [Online]. Available from: https://bahaiphilosophy.com/wp-content/uploads/2019/12/Drive_On_Through.pdf [Accessed 24 April 2020].

Koch, C. (2004). *The Quest for Consciousness: A Neurobiological Approach*. Englewood, CO: Roberts & Company Publishers.

Koch, C. (2009). When Does Consciousness Arise in Human Babies? *Scientific American*, 1 September. [Online]. Available from: https://www.scientificamerican.com/article/when-does-consciousness-arise/ [Accessed on 21 April 2020].

Koch, C. (2012a). *Consciousness: Confessions of a Romantic Reductionist*. Cambridge, MA: MIT Press.

Koch, C. (2012b). This is your brain on drugs. *Scientific American Mind*, 1 May. [Online]. Available from: http://www.scientificamerican.com/article/this-is-your-brain-on-drugs/ [Accessed 9 August 2016].

Koch, C. *et al.* (2016). Neural correlates of consciousness: progress and problems. *Nature Reviews Neuroscience*, 17: 307-321. DOI: https://doi.org/10.1038/nrn.2016.22.

Krioukov, D. *et al.* (2012). Network cosmology. *Scientific Reports*, 2, doi:10.1038/srep00793.

Kuhn, TS (2012). *The Structure of Scientific Revolutions*. Chicago, IL: The University of Chicago Press.

Kurzweil, R. (2005). *The Singularity Is Near*. New York, NY: Viking.

Landau, MJ *et al.* (2004). A Function of form: Terror management and structuring the social world. *Journal of Personality and Social Psychology*, 87 (2): 190-210.

Langer, E. and Rodin, J. (1976). The effects of choice and enhanced personal responsibility for the aged: a field experiment in an institutional setting. *Journal of Personality and Social Psychology*, 34 (2): 191-198.

Lapkiewicz, R. *et al.* (2011). Experimental non-classicality of an indivisible quantum system. *Nature*, 474: 490-493.

Laughlin, RB and Pines, D. (2000). The theory of everything. *Proceedings of the National Academy of Sciences of the United States of America*, 97 (1): 28-31. [Online]. Available from: https://www.pnas.org/content/97/1/28 [Accessed 26 April 2020].

Lee, KC *et al.* (2011). Entangling macroscopic diamonds at room temperature. *Science*, 334 (6060): 1253-1256.

Leggett, AN (2003). Nonlocal hidden-variable theories and quantum mechanics: An incompatibility theorem. *Foundations of Physics*, 33 (10): 1469-1493.

Levine, J. (1983). Materialism and qualia: The explanatory gap. *Pacific Philosophical Quarterly*, 64: 354-361.

Lewis, CR *et al.* (2017). Two dose investigation of the 5-HT-agonist psilocybin on relative and global cerebral blood flow. *NeuroImage*, July, doi:10.1016/j.neuroimage.2017.07.020.

Libet, B. (1985). Unconscious cerebral initiative and the role of conscious will in voluntary action. *Behavioral and Brain Sciences*, 8: 529-566.

Linde, A. (1998). *Universe, Life, Consciousness*. A paper delivered at the Physics and Cosmology Group of the "Science and Spiritual Quest" program of the Center for Theology and the Natural Sciences (CTNS), Berkeley, California. [Online]. Available from: web.stanford.edu/~alinde/SpirQuest.doc [Accessed 14 June 2016].

Lommel, P. van *et al.* (2001). Near-death experience in survivors of cardiac arrest: a prospective study in the Netherlands. *The Lancet*, 358 (9298): 2039-2045.

Luck, A. *et al.* (1999). Effects of video information on precolonoscopy anxiety and knowledge: a randomised trial. *The Lancet*, 354 (9195): 2032-2035.

Lynch, J. and Kilmartin, C. (2013). *Overcoming Masculine Depression: The Pain Behind the Mask*. New York, NY: Routledge.

Lythgoe, M. *et al.* (2005). Obsessive, prolific artistic output following subarachnoid hemorrhage. *Neurology*, 64 (2): 397-398.

Ma, X-S *et al.* (2013). Quantum erasure with causally disconnected choice. *Proceedings of the National Academy of Sciences of the Unites States of America*, 110 (4): 1221-1226.

Macnab, AJ *et al.* (2009). Asphyxial games or "the choking game": A potentially fatal risk behaviour. *Injury Prevention*, 15 (1): 45-49.

Maharaj, N. (1973). *I Am That*. Mumbai, India: Chetana Publishing.

Manning, AG *et al.* (2015). Wheeler's delayed-choice gedanken experiment with a single atom. *Nature Physics*, 11: 539-542.

Mathews, F. (2011). Panpsychism as paradigm. In: Blamauer, M. (ed.). *The Mental as Fundamental*. Frankfurt, Germany: Ontos Verlag.

Mead, GRS (translator) (2010). *The Corpus Hermeticum*. Whitefish, MT: Kessinger Publishing LLC.

Merali, Z. (2015). Quantum 'spookiness' passes toughest test yet. *Nature*, News, August 27. [Online]. Available from: http://www.nature.com/news/quantum-spookiness-passes-toughest-test-yet-1.18255 [Accessed 30 August 2015].

Merleau-Ponty, M. (1964). *The Primacy of Perception: And Other Essays on Phenomenological Psychology, the Philosophy of Art, History and Politics*. Evanston, IL: Northwestern University Press.

Miller, B. *et al.* (1998). Emergence of artistic talent in frontotemporal dementia. *Neurology*, 51 (4): 978-982.

Miller, B. *et al.* (2000). Functional correlates of musical and visual ability in frontotemporal dementia. *The British Journal of Psychiatry*, 176: 458-463.

Miller, G. (2005). What is the biological basis of consciousness? *Science*, 309 (5731): 79. DOI: 10.1126/science.309.5731.79.

Moore, T. (2012). *Care of the Soul: An Inspirational Programme to*

Add Depth and Meaning to Your Everyday Life. London, UK: Piatkus Books.

Moorjani, A. (2012). *Dying To Be Me: My Journey from Cancer, to Near Death, to True Healing*. Carlsbad, CA: Hay House.

Motl, L. (2009). Bohmists & segregation of primitive and contextual observables. *The Reference Frame,* 23 January. [Online]. Available from: https://motls.blogspot.com/2009/01/bohmists-segregation-of-primitive-and.html [Accessed 22 April 2020].

Muthukumaraswamy, SD *et al.* (2013). Broadband Cortical Desynchronization Underlies the Human Psychedelic State. *The Journal of Neuroscience,* 33 (38) 15171-15183. DOI: https://doi.org/10.1523/JNEUROSCI.2063-13.2013.

n.a. (2011). Signal for Consciousness in Brain Marked by Neural Dialogue. *Scientific American Mind,* 1 November. [Online]. Available from: https://www.scientificamerican.com/article/a-conversation-in-the-brain/ [Accessed 21 April 2020].

Nadeau, R. and Kafatos, M. (1999). *The Non-Local Universe: The new physics and matters of the mind*. Oxford, UK: Oxford University Press.

Nagasawa, Y. and Wager, K. (2016). Panpsychism and priority cosmopsychism. In: Brüntrup, G. and Jaskolla, L. (eds.). *Panpsychism*. Oxford, UK: Oxford University Press.

Nagel, T. (1974). What is it like to be a bat? *The Philosophical Review,* 83 (4): 435-450.

Nagel, T. (2012). *Mind and Cosmos: Why the Materialist Neo-Darwinian Conception of Nature Is Almost Certainly False*. Oxford, UK: Oxford University Press.

Narasimhan, A. and Kafatos, M. (2016). Wave Particle Duality, the Observer and Retrocausality. *arXiv:1608.06722 [quant-ph]*. [Online]. Available from: https://arxiv.org/abs/1608.06722 [Accessed 22 April 2020].

Neal, RM (2008). *The Path to Addiction: "And Other Troubles We Are Born to Know."* Bloomington, IN: AuthorHouse.

Neumann, J. von (1996). *Mathematical Foundations of Quantum Mechanics*. Princeton, NJ: Princeton University Press.

Newell, BR and Shanks, DR (2014). Unconscious influences on decision making: A critical review. *Behavioral and Brain Sciences*, 37 (1): 1-19.

Nixon, GM (2010). From panexperientialism to conscious experience: The continuum of experience. *Journal of Consciousness Exploration and Research*, 1 (3): 215-233.

Ockeloen-Korppi, CF *et al.* (2018). Stabilized entanglement of massive mechanical oscillators. *Nature*, 556: 478-482. DOI: https://doi.org/10.1038/s41586-018-0038-x.

O'Connell, AD *et al.* (2010). Quantum ground state and single-phonon control of a mechanical resonator. *Nature*, 464: 697-703. DOI: https://doi.org/10.1038/nature08967.

Okasha, S. (2002). *Philosophy of Science: A Very Short Introduction*. Oxford, UK: Oxford University Press.

Ortiz-Osés, A. (2008). *The Sense of the World*. Aurora, CO: The Davies Group Publishers.

Palhano-Fontes, F. *et al.* (2015). The psychedelic state induced by ayahuasca modulates the activity and connectivity of the default mode network. *PLoS ONE*, 10 (2): e0118143.

Paller, KA and Suzuki, S. (2014). The source of consciousness. *Trends in Cognitive Sciences*, 18 (8): 387-389.

Partington, CF (ed.) (1837). *The British Cyclopædia of Natural History: A Scientific Classification of Animals, Plants, and Minerals; With a Popular View of Their Habits, Economy, and Structure*, Vol. 3. London, UK: WS Orr & Co., Amen Corner, Paternoster-Row.

Pearl, J. (1988). *Probabilistic Reasoning in Intelligent Systems: Networks of Plausible Inference*. San Francisco, CA: Morgan Kaufmann.

Peres, J. *et al.* (2012). Neuroimaging during trance state: A contribution to the study of dissociation. *PLoS ONE*, 7 (11): e49360.

Petri, G. *et al.* (2014). Homological scaffolds of brain functional networks. *Journal of the Royal Society Interface*, 11 (101). DOI: https://doi.org/10.1098/rsif.2014.0873.

Piccinini, G. (2015). Computation in physical systems. In: Zalta, EN (ed.). *The Stanford Encyclopedia of Philosophy* (Summer 2015 Edition). [Online]. Available from: http://plato.stanford.edu/archives/sum2015/entries/computation-physicalsystems [Accessed 30 June 2016].

Piore, A. (2013). The genius within. *Popular Science*, March: 46-53.

Pollan, M. (2018). *How to Change Your Mind: What the New Science of Psychedelics Teaches Us About Consciousness, Dying, Addiction, Depression, and Transcendence*. New York, NY: Penguin.

Popper, K. (2005). *The Logic of Scientific Discovery*. London, UK: Routledge.

Priest, G. (2006). *In Contradiction: A study of the transconsistent*. Oxford, UK: Oxford University Press.

Proietti, M. (2019). Quantum physics: our study suggests objective reality doesn't exist. *The Conversation*, 14 November. [Online]. Available from: https://theconversation.com/quantum-physics-our-study-suggests-objective-reality-doesnt-exist-126805 [Accessed 23 April 2020].

Proietti, M. *et al.* (2019). Experimental rejection of observer-independence in the quantum world. *arXiv:1902.05080 [quant-ph]*. [Online]. Available from: https://arxiv.org/abs/1902.05080v1 [Accessed 22 April 2020].

Putnam, H. (1968). Is Logic Empirical? In: Cohen, RS and Wartofsky, MW (eds.). *Boston Studies in the Philosophy of Science*, 5. Dordrecht, Netherlands: D. Reidel, pp. 216-241.

Pyszczynski, T., Greenberg, J. and Solomon, S. (1997). Why do we need what we need? A terror management perspective on the roots of human social motivation. *Psychological Inquiry*, 8 (1): 1-20.

Rashkov, G. *et al.* (2019). Natural image reconstruction from brain waves: a novel visual BCI system with native feedback. *bioRxiv* preprint. DOI: https://doi.org/10.1101/787101.

Reddy, S. and Knight, K. (2011). What we know about the Voynich manuscript. In: *Proceedings of the 5th ACL-HLT Workshop on Language Technology for Cultural Heritage, Social Sciences, and Humanities*. Stroudsburg, PA: Association for Computational Linguistics, pp. 78-86.

Reiss, J. and Sprenger, J. (2017). Scientific objectivity. In: Zalta, EN (ed.). *The Stanford Encyclopedia of Philosophy* (Winter 2017 Edition). [Online]. Available from: https://plato.stanford.edu/entries/scientific-objectivity/ [Accessed 26 April 2020].

Retz (2007). Tripping without drugs: Experience with hyperventilation (ID 14651). *Erowid.org*. [Online]. Available from: http://www.erowid.org/exp/14651 [Accessed 25 February 2017].

Rhinewine, JP and Williams, OJ (2007). Holotropic breathwork: The potential role of a prolonged, voluntary hyperventilation procedure as an adjunct to psychotherapy. *The Journal of Alternative and Complementary Medicine*, 13 (7): 771-776.

Roberson, D. and Hanley, JR (2007). Color vision: Color categories vary with language after all. *Current Biology*, 17 (15): R605-R607. DOI: https://doi.org/10.1016/j.cub.2007.05.057.

Robinson, H. (2016). Dualism. In: Zalta, EN (ed.). *The Stanford Encyclopedia of Philosophy* (Spring 2016 Edition). [Online]. Available from: http://plato.stanford.edu/archives/spr2016/entries/dualism [Accessed 17 June 2016].

Romero, J. *et al.* (2010). Violation of Leggett inequalities in orbital angular momentum subspaces. *New Journal of Physics*, 12: 123007. [Online]. Available from: http://iopscience.iop.org/article/10.1088/1367-2630/12/12/123007 [Accessed 14 June 2016].

Ronnenberg, A. and Martin, K. (2010). *The Book of Symbols*. Cologne, Germany: Taschen.

Rosenberg, G. (2004). *A Place for Consciousness*. New York, NY:

Oxford University Press.

Rosner, RI, Lyddon, WJ and Freeman, A. (eds.) (2004). *Cognitive Therapy and Dreams*. New York, NY: Springer Publishing Company.

Ross, WD (1951). *Plato's Theory of Ideas*. Oxford, UK: Oxford University Press.

Rovelli, C. (1996). Relational quantum mechanics. *International Journal of Theoretical Physics*, 35 (8): 1637-1678.

Rovelli, C. (2018). *The Order of Time*. London, UK: Penguin.

Russell, B. (1995). *An Inquiry into Meaning and Truth*. London, UK: Routledge.

Russell, B. (2009). *Human Knowledge: Its Scope and Limits*. London, UK: Routledge Classics.

Ryle, G. (2009). *The Concept of Mind*. London, UK: Routledge.

Sacks, O. (1985). *The Man Who Mistook His Wife for a Hat*. New York, NY: Harper & Row.

Sandberg, A. and Boström, N. (2008). *Whole Brain Emulation: A Roadmap* (Technical Report #2008-3). Oxford, UK: Future of Humanity Institute, Oxford University. [Online]. Available from: http://www.fhi.ox.ac.uk/brain-emulation-roadmap-report.pdf [Accessed 26 February 2017].

Sartre, J-P (1992). *Being and Nothingness: A Phenomenological Essay on Ontology*. New York, NY: Washington Square Press.

Schaffer, J. (2010). Monism: The priority of the whole. *Philosophical Review*, 119 (1): 31-76.

Schartner, MM *et al.* (2017). Increased spontaneous MEG signal diversity for psychoactive doses of ketamine, LSD and psilocybin. *Scientific Reports*, 7, 46421. DOI: https://doi.org/10.1038/srep46421.

Schlumpf, Y. *et al.* (2014). Dissociative part-dependent resting-state activity in Dissociative Identity Disorder: A controlled fMRI perfusion study. *PloS ONE*, 9, doi:10.1371/journal.pone.0098795.

Schneider, S. (2018). Spacetime emergence, panpsychism and

the nature of consciousness: How does experience, which is so intimately tied to our perception of time and space, arise from timeless, non-spatial ingredients? *Scientific American*, 6 August. [Online]. Available from: https://blogs.scientificamerican.com/observations/spacetime-emergence-panpsychism-and-the-nature-of-consciousness/ [Accessed 23 April 2020].

Schooler, J. (2002). Re-representing consciousness: dissociations between experience and meta-consciousness. *Trends in Cognitive Science*, 6 (8): 339-344.

Seager, W. (2010). Panpsychism, aggregation and combinatorial infusion. *Mind and Matter*, 8 (2): 167-184.

Searle, JR (2004). *Mind: A Brief Introduction*. Oxford, UK: Oxford University Press.

Shani, I. (2015). Cosmopsychism: A holistic approach to the metaphysics of experience. *Philosophical Papers*, 44 (3): 389-437.

Shannon, CE (1948). A mathematical theory of communication. *Bell System Technical Journal*, 27: 379-423 & 623-656.

Shermer, M. (2011). What is pseudoscience? *Scientific American*, September 1. [Online]. Available from: http://www.scientificamerican.com/article.cfm?id=what-is-pseudoscience [Accessed 7 August 2016].

Siegel, E. (2016). Ask Ethan: Is the universe itself alive? *Forbes*, January 23. [Online]. Available from: http://www.forbes.com/sites/startswithabang/2016/01/23/ask-ethan-is-the-universe-itself-alive [Accessed 25 February 2017].

Skrbina, D. (2007). *Panpsychism in the West*. Cambridge, MA: MIT Press.

Smolin, L. (2007). *The Trouble with Physics: The Rise of String Theory, the Fall of a Science, and What Comes Next*. New York, NY: Mariner Books.

Smolin, L. (2013). *Time Reborn: From the Crisis in Physics to the Future of the Universe*. Boston, MA: Houghton Mifflin

Harcourt.

Stannard, DE (1980). *Shrinking History: On Freud and the Failure of Psychohistory*. Oxford, UK: Oxford University Press.

Stapp, HP (1993). *Mind, Matter, and Quantum Mechanics*. New York, NY: Springer-Verlag.

Stapp, HP (2001). Quantum Theory and the Role of Mind in Nature. *Foundations of Physics*, 31: 1465-1499. DOI: https://doi.org/10.1023/A:1012682413597.

Stapp, HP (2007). *Mindful Universe: Quantum mechanics and the participating observer*. New York, NY: Springer-Verlag.

Stapp, HP (2017). *Quantum Theory and Free Will: How mental intentions translate into bodily actions*. New York, NY: Springer-Verlag.

Stavrova, O., Ehlebracht, D. and Fetchenhauer, D. (2016). Belief in scientific-technological progress and life satisfaction: The role of personal control. *Personality and Individual Differences*, 96: 227-236.

Stoerig, P. and Cowey, A. (1997). Blindsight in man and monkey. *Brain*, 120 (3): 535-559.

Stoljar, D. (2016). Materialism. In: Zalta, EN (ed.). *The Stanford Encyclopedia of Philosophy* (Spring 2016 Edition). [Online]. Available from: http://plato.stanford.edu/archives/spr2016/entries/materialism/ [Accessed 26 February 2017].

Strasburger, H. and Waldvogel, B. (2015). Sight and blindness in the same person: Gating in the visual system. *PsyCh Journal*, 4 (4): 178-185.

Strassman, R. (2001). *DMT: The Spirit Molecule*. Rochester, VT: Park Street Press.

Strassman, R. *et al.* (2008). *Inner Paths to Outer Space*. Rochester, VT: Park Street Press.

Strawson, G. *et al.* (2006). *Consciousness and Its Place in Nature*. Exeter, UK: Imprint Academic.

Strawson, G. (2013). Real Naturalism. *London Review of Books*, 26 September, 35 (18). [Online]. Available from: https://www.

lrb.co.uk/the-paper/v35/n18/galen-strawson/real-naturalism [Accessed 26 April 2020].

Strawson, G. (2018). The consciousness deniers. *The New York Review of Books*, 13 March. [Online]. Available from: https://www.nybooks.com/daily/2018/03/13/the-consciousness-deniers/ [Accessed 25 April 2020].

Streater, RF (2007). *Lost Causes in and beyond Physics*. Berlin, Germany: Springer-Verlag.

Swanson, LR (2018). Unifying theories of psychedelic drug effects. *Frontiers in Pharmacology*, 2 March. DOI: https://doi.org/10.3389/fphar.2018.00172.

Swedenborg, E. (2007). *Heaven and its Wonders and Hell*. Charleston, SC: BiblioBazaar.

Tagliazucchi, E. *et al.* (2014). Enhanced repertoire of brain dynamical states during the psychedelic experience. *Human Brain Mapping*, 35 (11): 5442-5456. DOI: https://doi.org/10.1002/hbm.22562.

Tan, LH *et al.* (2008). Language affects patterns of brain activation associated with perceptual decision. *Proceedings of the National Academy of Sciences of the United States of America*, 105 (10): 4004-4009. DOI: https://doi.org/10.1073/pnas.0800055105.

Tarlaci, S. and Pregnolato, M. (2016). Quantum neurophysics: From non-living matter to quantum neurobiology and psychopathology. *International Journal of Psychophysiology*, 103: 161-173.

Tarnas, R. (2010). *The Passion of the Western Mind*. London, UK: Pimlico.

Taylor, C. (2007). *A Secular Age*. Cambridge, MA: Harvard University Press.

Taylor, JB (2009). *My Stroke of Insight: A Brain Scientist's Personal Journey*. New York, NY: Viking.

Taylor, K. (1994). *The Breathwork Experience: Exploration and Healing in Nonordinary States of Consciousness*. Santa Cruz, CA: Hanford Mead.

Tegmark, M. (2014). *Our Mathematical Universe: My Quest for the Ultimate Nature of Reality*. New York, NY: Vintage Books.

Tillich, P. (1952). *The Courage To Be*. New Haven, CT: Yale University Press.

Timmermann, C. *et al.* (2018). DMT models the near-death experience. *Frontiers in Psychology*, 15 August. DOI: https://doi.org/10.3389/fpsyg.2018.01424.

Tittel, W. *et al.* (1998). Violation of Bell inequalities by photons more than 10 km apart. *Physical Review Letters*, 81 (17): 3563-3566.

Tongeren, DR van and Green, JD (2010). Combating meaninglessness: On the automatic defense of meaning. *Personality and Social Psychology Bulletin*, 36 (10): 1372-1384.

Tononi, G. (2004). An information integration theory of consciousness. *BMC Neuroscience*, 5 (42), doi: 10.1186/1471-2202-5-42.

Treffert, D. (2006). *Extraordinary People: Understanding Savant Syndrome*. Omaha, NE: iUniverse, Inc.

Treffert, D. (2009). The savant syndrome: An extraordinary condition. A synopsis: Past, present, future. *Philosophical Transactions of the Royal Society B*, 364 (1522): 1351-1357.

Tsuchiya, N. *et al.* (2015). No-report paradigms: Extracting the true neural correlates of consciousness. *Trends in Cognitive Science*, 19 (12): 757-770.

Urgesi, C. *et al.* (2010). The spiritual brain: Selective cortical lesions modulate human self-transcendence. *Neuron*, 65: 309-319.

Vaillant, GE (1992). *Ego Mechanisms of Defense: A Guide for Clinicians and Researchers*. Washington, DC: American Psychiatric Press.

Valsiner, J. (1998). *The Guided Mind*. Cambridge, MA: Harvard University Press.

Vandenbroucke, A. *et al.* (2014). Seeing without knowing: Neural signatures of perceptual inference in the absence of report.

Journal of Cognitive Neuroscience, 26 (5): 955-969.

Varela, FJ, Thompson, E. and Rosch, E. (1993). *The Embodied Mind: Cognitive Science and Human Experience*. Cambridge, MA: MIT Press.

Vazza, F. and Feletti, A. (2017). The strange similarity of neuron and galaxy networks: Your life's memories could, in principle, be stored in the universe's structure. *Nautilus*, July 20. [Online]. Available from: http://nautil.us/issue/50/emergence/the-strange-similarity-of-neuron-and-galaxy-networks [Accessed 21 July 2017].

Walls, LD (2003). *Emerson's Life in Science: The Culture of Truth*. Ithaca, NY: Cornell University Press.

Watts, A. (1989). *The Book: On the Taboo Against Knowing Who You Are*. New York, NY: Vintage Books.

Webster, R. (1995). *Why Freud Was Wrong: Sin, Science, and Psychoanalysis*. New York, NY: Basic Books.

Wegner, DM (2002). *The Illusion of Conscious Will*. Cambridge, MA: MIT Press.

Weihs, G. *et al.* (1998). Violation of Bell's inequality under strict Einstein locality conditions. *Physical Review Letters*, 81 (23): 5039-5043.

Westen, D. (1999). The scientific status of unconscious processes: Is Freud really dead? *Journal of the American Psychoanalytic Association*, 47 (4): 1061-1106.

Whinnery, J. and Whinnery, A. (1990). Acceleration-induced loss of consciousness. A review of 500 episodes. *Archives of Neurology*, 47 (7): 764-776.

Whitehead, AN (1947). *Essays in Science and Philosophy*. New York, NY: Philosophical Library.

Wigner, E. (1960). The unreasonable effectiveness of mathematics in the natural sciences. *Communications on Pure and Applied Mathematics*, 13 (1): 1-14.

Windt, JM and Metzinger, T. (2007). The philosophy of dreaming and self-consciousness: what happens to the experiential subject during the dream state? In: Barrett, D. and McNamara, P. (eds.). *The New Science of Dreaming*. Westport, CT: Praeger, pp. 193-247.

Windt, J., Nielsen, T. and Thompson, E. (2016). Does consciousness disappear in dreamless sleep? *Trends in Cognitive Sciences*, 20 (12): 871-882.

Wolchover, N. (2018). Famous experiment dooms alternative to quantum weirdness. *Quanta Magazine*, 11 October. [Online]. Available from: https://www.quantamagazine.org/famous-experiment-dooms-pilot-wave-alternative-to-quantum-weirdness-20181011/ [Accessed 22 April 2020].

Yetter-Chappell, H. (2018). Idealism without God. In: Goldschmidt, T. and Pearce, K. (eds.). *Idealism: New Essays in Metaphysics*. Oxford, UK: Oxford University Press.

Zemach, E. (2006). Wittgenstein's philosophy of the mystical. In: Copi, IM and Beard, RW (eds.). *Essays on Wittgenstein's Tractatus*. London, UK: Routledge, pp. 359-376.

Zicheng, H. (2006). *Vegetable Roots Discourse: Wisdom from Ming China on Life and Living: Caigentan*. Berkeley, CA: Shoemaker & Hoard.

Zurek, WH (1994). Preferred Observables, Predictability, Classicality, and the Environment-Induced Decoherence. *arXiv:gr-qc/9402011*. [Online]. Available from: https://arxiv.org/abs/gr-qc/9402011 [Accessed 22 April 2020].

Zurek, WH (2003). Decoherence and the transition from quantum to classical – REVISITED. *arXiv:quant-ph/0306072*. [Online]. Available from: https://arxiv.org/abs/quant-ph/0306072 [Accessed 22 April 2020].

ACADEMIC AND SPECIALIST

Iff Books publishes non-fiction. It aims to work with authors and titles that augment our understanding of the human condition, society and civilisation, and the world or universe in which we live.
If you have enjoyed this book, why not tell other readers by posting a review on your preferred book site.

Recent bestsellers from Iff Books are:

Why Materialism Is Baloney

How true skeptics know there is no death and fathom answers to life, the universe, and everything

Bernardo Kastrup

A hard-nosed, logical, and skeptic non-materialist metaphysics, according to which the body is in mind, not mind in the body.

Paperback: 978-1-78279-362-5 ebook: 978-1-78279-361-8

The Fall

Steve Taylor

The Fall discusses human achievement versus the issues of war, patriarchy and social inequality.

Paperback: 978-1-78535-804-3 ebook: 978-1-78535-805-0

Brief Peeks Beyond

Critical essays on metaphysics, neuroscience, free will, skepticism and culture

Bernardo Kastrup

An incisive, original, compelling alternative to current mainstream cultural views and assumptions.

Paperback: 978-1-78535-018-4 ebook: 978-1-78535-019-1

Framespotting

Changing how you look at things changes how you see them

Laurence & Alison Matthews

A punchy, upbeat guide to framespotting. Spot deceptions and hidden assumptions; swap growth for growing up. See and be free.

Paperback: 978-1-78279-689-3 ebook: 978-1-78279-822-4

Is There an Afterlife?

David Fontana

Is there an Afterlife? If so what is it like? How do Western ideas of the afterlife compare with Eastern? David Fontana presents the historical and contemporary evidence for survival of physical death.

Paperback: 978-1-90381-690-5

Nothing Matters

a book about nothing

Ronald Green

Thinking about Nothing opens the world to everything by illuminating new angles to old problems and stimulating new ways of thinking.

Paperback: 978-1-84694-707-0 ebook: 978-1-78099-016-3

Panpsychism

The Philosophy of the Sensuous Cosmos

Peter Ells

Are free will and mind chimeras? This book, anti-materialistic but respecting science, answers: No! Mind is foundational to all existence.

Paperback: 978-1-84694-505-2 ebook: 978-1-78099-018-7

Punk Science

Inside the Mind of God

Manjir Samanta-Laughton

Many have experienced unexplainable phenomena; God, psychic abilities, extraordinary healing and angelic encounters. Can cutting-edge science actually explain phenomena previously thought of as 'paranormal'?

Paperback: 978-1-90504-793-2

The Vagabond Spirit of Poetry

Edward Clarke

Spend time with the wisest poets of the modern age and of the past, and let Edward Clarke remind you of the importance of poetry in our industrialized world.

Paperback: 978-1-78279-370-0 ebook: 978-1-78279-369-4

Readers of ebooks can buy or view any of these bestsellers by clicking on the live link in the title. Most titles are published in paperback and as an ebook. Paperbacks are available in traditional bookshops. Both print and ebook formats are available online. Find more titles and sign up to our readers' newsletter at http://www.johnhuntpublishing.com/non-fiction

Follow us on Facebook at https://www.facebook.com/JHPNonFiction

and Twitter at https://twitter.com/JHPNonFiction